MW01504519

American University Studies

Series XIV
Education

Vol. 6

PETER LANG
New York · Berne · Frankfort on the Main · Nancy

Jeffrey Kane

Beyond Empiricism

Michael Polanyi Reconsidered

JEROME LIBRARY–BOWLING GREEN STATE UNIVERSITY

PETER LANG
New York · Berne · Frankfort on the Main · Nancy

CIP-Kurztitelaufnahme der Deutschen Bibliothek

Kane, Jeffrey:
Beyond empiricism: Michael Polanyi reconsidered /
Jeffrey Kane. – New York; Berne; Frankfort on
the Main; Nancy: Lang, 1984.
    (American University Studies: Ser. 14,
    Education; Vol. 6)
    ISBN 0-8204-0118-8

NE: American University Studies / 14

Library of Congress Catalog Card Number:
84-47540
ISBN 0-8204-0118-8
ISSN 0740-4565

© Peter Lang Publishing, Inc., New York 1984

All rights reserved.
Reprint or reproduction, even partially, in all forms such as
microfilm, xerography, microfiche, microcard, offset prohibited.

Printed by Lang Druck, Inc., Liebefeld/Berne (Switzerland)

FOR JANET, GABRIEL, AND EMILY

TABLE OF CONTENTS

INTRODUCTION

Michael Polanyi claims that epistemologists from Hume to Karl Popper have considered only the articulate, impersonal and explicit components of thought and, specifically, scientific thought. The Humean epistemological legacy is based upon the idea that there is no logical justification for the belief that inferences drawn from past experiences will necessarily be consistent with future events. Popper has offered what he believes is a logically consistent framework for the growth of scientific knowledge. He agrees with Hume that there are no justifications for inferential knowledge, but he maintains that by transposing observations and theories into logical terms, non-justifiable statements can be refuted.

Polanyi does not attempt to restore or revitalize the concept of induction but rather to create a philosophical concept of knowledge based upon what may be called an existential or phenomenological relation between the knower and the known. He rejects the formalistic concept of knowledge underlying the Humean dilemma and Popper's non-justificationist solution. Polanyi rejects the ideal of detached and impersonal knowledge and replaces it with a personal model of knowledge which requires tacit, informal activity on the part of the individual knower. Polanyi writes,

> Modern man has set up as the ideal of knowledge the conception of science as a set of statements which is "objective" in the sense that its substance is entirely determined by observation, even while its presentation may be shapen by convention. This conception would be shattered if the intuition of rationality in nature had to be acknowledged as a justifiable and indeed essential part of scientific theory.[1]

One can readily see that Polanyi's orientation could run the risk of diffusing the pursuit of scientific knowledge into the random sifting of inclination according to subjective and arbitrary standards. Polanyi compounds the possibility of such a dilemma with his interdisciplinary mode of argumentation.

The purpose of this book is to systematically clarify and elucidate Polanyi's epistemology. More specifically, it will examine Polanyi's notion of the "tacit," intuitive, and imaginative capacities of scientists as they proceed with their scientific discoveries and assessments. A clear exposition of these heuristic aspects of Polanyi's epistemology could affect the standards and structures of both the scientific and educational communities.

---

[1]Michael Polanyi, _Personal Knowledge Towards a Post-Critical Philosophy_ (New York: Harper & Row, Inc., 1962), p. 16.

Chapter I introduces the reader to Polanyi's metaphysic. His notion of the ontological referents of science is critically analyzed to identify the logical structure of objectivity. Positivistic and non-justificationistic conceptions of objectivity and, corollary, scientific inquiry, are rejected. A basis for a heuristic concept of knowledge is provided.

In Chapter II, Polanyi's metaphysic is shown to preclude the possibility of a completely logical or empirically determined process of scientific discovery or assessment. The primary role of indeterminate personal intellectual factors and the secondary role of explicit procedures and methods are analyzed.

Chapter III elaborates the nature and structure of the indeterminate aspects of the process of knowing. It focuses upon Polanyi's concept of "indwelling." The relation of these tacit elements of knowledge and articulate knowledge is structurally outlined. After much analysis, Polanyi's concept of mind is found to be problematic.

Chapter IV draws on the analysis of the three previous chapters to construct a detailed model of scientific inquiry and discovery. The nature of and dynamic interaction of Polanyi's concepts of "intuition" and "imagination" are examined. Polanyi's concepts are extended and new functional relationships are advanced where Polanyi does not provide sufficient detail.

Chapter V summarizes the preceding four chapters to give the reader an overview of the arguments presented. Implications of the elements of Polanyi's "personal" model of knowledge are drawn for the general scientific and educational communities.

CHAPTER I

METAPHYSICAL OBJECTIVITY

Michael Polanyi believes that the concept of impersonal objectivity that, in varying forms, has under-girded modern science, is misguided. He maintains that there is a necessary personal component to the conduct of scientific inquiry, and that it is this component that propels science forward.

In order to understand Polanyi's position, it is most appropriate to begin with a study of his concept of the proper referents of scientific research. His metaphysic will then provide us a focus for the establishment of a general model of personal knowing and ultimately of a unique concept of objectivity.

Polanyi explains his initial position by stating,

> To say that the discovery of objective truth in science consists in the apprehension of a rationality which commands our respect and arouses our contemplative admiration; that such discovery, while using the experience of our senses as clues, transcends this experience by embracing the vision of a reality beyond the impressions of our senses, a vision which speaks for itself in guiding us to an ever deeper understanding of reality--such an account of scientific procedure would be generally shrugged aside as outdated Platonism: a piece of mystery-mongering unworthy of an enlightened age.

> Yet it is precisely on this conception of
> objectivity that I wish to insist....[1]

Polanyi conceives of scientific inquiry as an attempt to make natural occurrences rationally comprehensible. He rejects the positivistic notion of science which maintains that scientific theory is a convenient summary of experience wholly delimited by empirical observation. He also rejects Karl Popper's relegation of science to the "third world," a world of rational thought whose referents include only purely explicit, logical test statements and explanatory theories. The purpose of scientific inquiry, according to Polanyi, "is to discover the hidden reality underlying the facts of nature."[2] Thus, the referents of his concept of scientific theory transcend the limitations of empirical observations and include implications that extend beyond the logical principles governing the entities in Popper's third world.

The question arises as to what is meant by the term "reality" when it is used to denote something which lies beyond direct empirical experience. Polanyi contends that the meaning of "reality" within this context is

_____

[1]Michael Polanyi, Personal Knowledge: Towards a Post-Critical Philosophy (New York: Harper & Row, 1964), pp. 5 and 6.

[2]Michael Polanyi, "Science and Reality," The British Journal for the Philosophy of Science 18 (November 1967): 177.

evident in the Copernican Revolution. "We do not see it recognized," writes Polanyi, "that in the way Copernicus interpreted [his] discovery, he and his followers established the metaphysical grounds of modern science."[3]

Copernicus, Polanyi maintains, rejected Ptolemaic anthropocentrism, not because of its failure to accurately calculate the movements of the planets,[4] but because it did not disclose the mathematical foundations of the universe transcending the parameters of observations made from the earth alone. Copernicus was not content with calculation but was intent upon discovering the mathematical determinants of the motions of the planets.

Ptolemaic theory had referred to celestial motions from the singular perspective of an observer on earth and attempted to provide an interpretative framework solely for such an observer. Polanyi notes that within such a system--being closely related to the modern positivistic philosophy of science--"it does not matter whether [hypotheses] are true or false, provided that they reproduce exactly the phenomena of nature."[5]

Copernicus, however, wanted to find a mathematical

---

[3]Ibid.

[4]Polanyi notes that both Copernicus and Ptolemaic theory could account for all observations made at the time of and for 148 years subsequent to the publication of De Revolutionibus. Personal Knowledge, p. 152.

[5]Polanyi, Personal Knowledge, p. 146.

formulation which consistently accounted for observations of celestial phenomena from any and all terrestrial locations; to discover something that was ontologically accurate and therefore true. Polanyi concludes that the claim of greater objectivity for Copernican theory was justified not for its superior capacity to calculate celestial activities, but for its rational structure which could account for such activities as observed from an infinite number of locations.[6]

In order to form such a conception of celestial law, Copernicus had to disregard much of the empirical information available to him. E.A. Burtt, in The Metaphysical Foundations of Modern Science, states that

> it is safe to say that even had there been no religious scruples whatever against the Copernican astronomy, sensible men all over Europe, especially the most empirically minded, would have pronounced it a wild appeal to accept the premature fruits of an uncontrolled imagination....Contemporary empiricists, had they lived in the sixteenth century, would have been first to scoff out of court the new philosophy of the universe.[7]

Polanyi explains that when observed from the earth, the individual planets move at varying speeds and follow looping rather than linear west to east paths. The rhythmical patterns of the movements of the planets were

---

[6]Polanyi, Personal Knowledge, p. 5.

[7]Edwin Arthur Burtt, The Metaphysical Foundations of Modern Science (Garden City: Doubleday & Company, Inc., 1932), p. 38.

actual in the sense that they could be observed. Had
Copernicus merely been concerned with empirical evidence
and limited his investigation to such considerations, he
would have been satisfied with a mathematical formulation accounting for them, a mathematical formulation like
Ptolemaic theory. Copernicus, however, was interested in
transcending appearances to rationally understand the
actual movement of the planets. Such a comprehensive
framework would, Copernicus believed, unite in a single
system numerous apparently unrelated phenomena. The
specifics of the problems confronting Copernicus are beyond the scope of this thesis, but the following quotation gives a succinct indication of his search for
rationality in celestial events rather than a mathematical system designed to manage the given data.

> A striking example of this in the case
> of Copernicus is provided by the way in
> which his theory explained the fact that
> all the Ptolemaic major epicycles were
> of one terrestrial year's duration.
> Ptolemy was himself aware of this, but
> he accorded it no significance, since
> there was no reason in his system why
> epicycles should be of any particular size,
> and it would not have materially affected
> anything if one or the other were different.
> This noteworthy correlation was passed off
> as coincidence, just as we now pass off
> equally strong correlations. For the
> Copernican system, of course, not only did
> this make sense, but it was a necessary
> consequence. Further, because their appar-
> ent oscillations were of different ampli-
> tudes, Copernicus was able to predict the
> sequence of the orbital radii, and it was
> this later achievement with which he him-
> self was most impressed because it showed

that he had discovered an order, a coherence in nature.[8]

Copernicus succeeded in making the movement of the planets part of a rational intellectual framework firmly grounded in Pythagorean notions, a framework which not only accurately described and predicted the available celestial motions but which integrated them into a cohesive unity. Celestial reality, for Copernicus, was not circumscribed by empirical information but rather included those principles or laws accessible to him as a rational thinker.[9] The scientific theory he developed was intended to refer to the actual foundations of celestial motion as they could be rationally comprehended in accordance with Pythagorean notions.[10] In short, he sought to conceptualize the laws ordering appearances and not merely to "establish the functional relations of

---

[8] John Brennan, "Polanyi's Transcendence of the Distinction Between Objectivity and Subjectivity as Applied to the Philosophy of Science," Journal of the British Society for Phenomenology 8, 3 (October 1977):148.

[9] This idea was dependent upon Copernicus's assumption of Pythagorean metaphysics. Copernican rationality was based upon Pythagorean presuppositions.

[10] It is important to note that ontological reference is a matter of intention rather than necessarily a matter of actualization. Polanyi would argue, for example, that while the scientific community continued to maintain the rationality of a heliocentric depiction of the solar system, Kepler eventually disregarded the Copernican notion of the circularity of planetary orbits. The point is that ontological intentions include the ever present possibility of error.

sensory data observed by our senses."[11] Polanyi illustrates Copernicus' quest for the rational laws of celestial motion with the following quotation from Copernicus in which the latter claims that he has

> at last discovered that, if the motions of the rest of the planets be brought into relation with the circulation of the Earth and be reckoned in proportion to the orbit of each planet, not only do their phenomenon presently ensue, but the orders and magnitudes of all stars and spheres, nay the heavens themselves, become so bound together that nothing in any part thereof could be moved from its place without producing confusion in all other parts of the Universe as a whole.[12]

"The Universe as a whole" is here a rational concept referring to an order and lawfulness transcending the limitations of sensory experience. "Everything is now bound together, he claims, and this is a sign that the *system is real*."[13] The looping paths of the planets are now understood as "mere illusions, without any underlying epicyclical motions."[14]

The superior objectivity of Copernican theory was related to its transcendence of observed phenomena rather than its sole dependence upon them. Scientific theory

---

[11]Polanyi, The British Journal for the Philosophy of Science, p. 178.

[12]Ibid., p. 185.

[13]Ibid.

[14]Ibid., p. 121.

was thus believed to refer to aspects of reality inti-
mated but not fully revealed by sensory evidence.
Polanyi writes of Copernicus, "Thus, did he make the
metaphysical claim that science can discover new knowledge about fundamental reality and thus did this claim
eventually triumph in the Copernican revolution."[15]

The reality to which theories thus apply extends
beyond that which is observed. That which is perceived
is, to a degree, non-objective, or merely partial.
Polanyi states that, "instead of being the clear and firm
ground underlying all appearances [reality] will turn
out to be known only vaguely, with an unlimited range
of unspecifiable expectations attached to it."[16]

Polanyi's concept of the metaphysical referents of
science is paralleled in the structure of scientific
knowledge. As science may be said to refer to ontological entities so, Polanyi maintains, scientific knowledge
is based upon intellectual tenets that transcend given
theories and explicit procedures. He asserts that all
scientific inquiry is rooted in fiduciary commitments
that can neither be justified nor refuted in a formal
manner.

We can best understand the "fiduciary-rootedness"

---

[15]Polanyi, _The British Journal for the Philosophy of Science_, p. 187.

[16]Ibid., p. 177.

of scientific knowledge through a comparison of Polanyi's treatment with that of Karl Popper with regard to what has historically been known as the "Tu Quoque Argument." The argument runs that "rationalism rests on an irrational choice of and commitment to rationality [and that] rationalism is as irrational as any other commitment.[17]

Popper explains,

> Neither logical argument nor experience can establish the rationalist attitude; for only those who are ready to consider argument or experience, and who have therefore adopted this attitude already, will be impressed by them....No rational argument will have a rational effect on a man who does not want to adopt a rational attitude.[18]

Popper maintains that we must reject what he calls an "uncritical" or "comprehensive" rationalism that attempts to establish rationalism by argument or recourse to experience; he does not believe, however, that the only alternative is to commit oneself to irrationalism. Popper commits himself to what he calls a "critical" rationalism that recognizes the limitations of reason to justify its own existence. Critical rationalism admits its origins lie in a "faith in reason."

---

[17]Joseph Agassi, "Rationality and the 'Tu Quoque Argument,'" Inquiry 16 (Winter 1973):395.

[18]Karl R. Popper. The Open Society and Its Enemies (Princeton, NJ: Princeton University Press, 1966), Vol. II, pp. 230-231.

The critical issue for our present concern is that Popper maintains that this single leap of faith is the only one necessary to begin scientific inquiry and to carry out scientific assessment.

We find this contention reflected in Popper's treatment of the Humean question of the possibility of a justification for inductive inferences. He maintains that the process of induction by repetition can in no way be justified. "The idea of induction by repetition must be due to an error--a kind of optical illusion. In brief: there is no such thing as induction by repetition."[19] Popper's argument that rationalism (critical rationalism) is grounded in an act of faith precludes the possibility of induction: the concept of induction necessarily implies that rational or systematic knowledge can be derived from empirical observation but the intention of arriving at such knowledge must begin with a non-empirically derived commitment to it.

This is not to say that Popper expressly presents such a response to the Humean problem of induction. Rather, it is to say that he thus establishes a foundation whereby he is not, like many others, forced by Hume to retreat to irrationalism. He sees Hume's own

---

[19]Karl R. Popper, Objective Knowledge (Oxford: The Clarendon Press, 1972), pp. 6, 7.

formulations as providing the basis for an irrationalist epistemology. He states that Hume's

> result that repetition has no power whatever
> as an argument, although it dominates our
> cognitive life or our "understanding," led
> him to the conclusion that argument or
> reason plays only a minor role in our under-
> standing. Our knowledge "is unmasked as
> being not only of the nature of belief, but
> of rationally indefensible belief--of an
> irrationalist faith."[20]

Popper's critical rationalism sets the stage for him to restate Hume's dilemma in such a way as to minimize the initial leap of faith underlying such rationalism and to erect a system of knowledge without any further reference to irrational components. Thus, when Popper focuses upon Hume's question of whether or not we can be "justified in reasoning from (repeated) instances of which we have experience to other instances (conclusions) of which we have no experience," he restates it in logical terms. By transcribing observable events into "test statements" and conclusions in regard to other natural events into "explanatory universal theories" Popper attempts to eliminate any and all questions of belief or faith from questions of knowledge assessment.

Once Popper restates the problem of induction in logical terms, he maintains that the critical question is

---

[20]Popper, Objective Knowledge, p. 4.

"...can the assumption of the truth of test statements justify either the claim that a universal theory is true or the claim that it is false."[21] Popper's answer is that we can be justified in holding a universal theory as false where "empirical evidence" refutes it. It would be irrational to hold that a universal statement was true if it had been refuted.

Thus, Popper concentrates his effort on restating the problem of induction so as to allow for the development of universal assertions, assertions capable of purely impersonal and objective assessment. His logical restatement of the problem of induction, in making a single leap to rationality, transforms the object of knowledge, and specifically scientific knowledge, from subjective impressions to logical processes and structures. Having embedded the object of science in such a logical ground, the relation of the various elements could be critically reviewed. Popper holds that the proper referents of science are not to be found either in an ontological or personal domain but in a "third world" of logically and impersonally related assertions that may or may not one day be refuted. It is, he maintains, a "world of objective theories, objective problems, and objective arguments."[22]

_____

[21]Popper, Objective Knowledge, p. 7.

[22]Ibid., p. 108.

The question of whether or not one <u>believes</u> in or <u>doubts</u> a given theory is, according to Popper, of no consequence because knowledge in the third world is "totally independent of anybody's claim to know....Knowledge in the objective sense is <u>knowledge without a knower</u>: it is knowledge without a knowing subject."[23] Popper hopes to overcome the impotence of logic as it is enmeshed with psychology, as found in the works of Hume and what Popper calls "belief Philosophers," by providing a unitary rational framework for the assessment of knowledge claims. This is not to say that knowledge claims can be justified, but rather that competing knowledge claims, can to a degree,[24] be objectively assessed and compared.

The specific theoretical and practical formulations for assessing the relative value claims of differing theories is not significant for us at this point. What is critical for our purpose (and that is ultimately to clarify the basic tenets of Polanyi's concept of objectivity) is that Popper's treatment of both the Tu Quoque Argument and Hume's problem attempts to develop a totally explicit and rational concept of objectivity.

[23]Popper, <u>Objective Knowledge</u>, p. 109.

[24]The notion of degree in this case indicates that we cannot be justified in holding that a scientific assertion is true but only that one is false. No amount of positive cases (repetitions) can justify a knowledge claim.

Polanyi, unlike Popper, who attempts to minimize the a-critical premises of science by making a singular leap to rationality, suggests that numerous premises assimilated in the process of our intellectual development, a-critically and habitually enter into the conduct of inquiry. Joseph Agassi, in comparing Popper's and Polanyi's treatment of the question of the foundations of rationality, concludes that whereas "Popper says he leaps to rationality, Polanyi says [he leaps] to science."[25] In other words, he does not make a single leap to a rational plateau of assessment but argues that presuppositions play into all levels and all kinds of scientific activities. He thus develops a model of scientific inquiry which is consistent with his concept of objectivity in that at a generative but non-specific sub-structure, a personal participation of the knower in the known.

Polanyi stresses that the content of the mind is not wholly explicit, that scientists do not merely (cannot merely) commit themselves solely to rational principles. He maintains that various conceptual and even perceptual acts rely upon numerous and interlocking premises. Polanyi writes,

> The standards of intellectual satisfaction
> which urge and guide our eyes to gather what
> there is to see, and which guide our thoughts

---

[25]Agassi, Inquiry, p. 397.

also shape our conception of things--the
beliefs about the nature of things trans-
mitted by our everyday descriptive langu-
age--all these form part of the premise
of science, even though we must allow for
the revision of these standards and beliefs
within science.[26]

Here lies Polanyi's response to the Tu Quoque

Argument: he maintains that scientists, as well as indi-

viduals engaged in other intellectual pursuits, a-critic-

ally (irrationally) commit themselves to certain intel-

lectual premises. The scientist's cognitive framework,

for Polanyi, therefore includes not only rational con-

cepts and processes but also unformalized notions of the

nature of reality and of science. Even Popper's concep-

tion of objective scientific knowledge presumes that the

goal of science is the elimination of personal components

in knowledge assessment, a goal Polyani believes to be

misguided because he asserts that any and all intellec-

tual pursuits incorporate a variety of a-critical and

informal assumptions and processes.

We may liken these factors to Thomas Kuhn's notion

of the tenets of tradition that course through the para-

digms employed in science. Kuhn maintains that the prem-

ises of science are indefinite and loosely bound but

quite capable of shaping the nature and scope of scien-

tific inquiry. He explains that the research problems

---

[26]Polanyi, Personal Knowledge, p. 161.

and techniques of normal science are not bound by

> some explicit or even some fully-discover-
> able set of rules and assumptions...instead
> they may relate by resemblance and by model-
> ing to one or another part of the scientific
> corpus which the community in question
> already recognizes as among its established
> achievements.[27]

It might be argued that Kuhn is referring to the way

science has evolved rather than how it should proceed;

the belief that science has progressed through the in-

corporation of central paradigms does not justify or ex-

plain the notion that such paradigmatic structures are

necessary. Here, Polanyi responds by indicating the

nature and function of the premises of science.

He notes that facts do not speak for themselves.

Baconian ideals have long been recognized as functionally

and philosophically inadequate. Scientists, according to

Polanyi, do not merely look _at_ things but _for_ things, for

those things judged to be significant for some reason.

The scientist necessarily selects what he believes to be

significant areas of inquiry, effective methods, effi-

cient measures, acceptable scales of probability, etc.

(The fact that certain standards and methods can be

employed to cut down the personal responsibility of the

scientist only serves to strengthen the point that

---

[27]Thomas S. Kuhn, _The Structure of Scientific
Revolutions_ (Chicago, IL: The University of Chicago Press,
1970), pp. 45, 46.

scientists operate within a socio-cultural field which includes basic and a-critical premises.) In short, Polanyi maintains that the premises of science are established before the fact and employed to accredit the "factuality" of evidence.

> We do not believe in the existence of facts
> because of our anterior and securer belief
> in any explicit logical presuppositions of
> such a belief; but on the contrary, we be-
> lieve in certain explicit presuppositions
> of factuality only because we have discovered
> that they are implied in our belief in the
> existence of fact.[28]

Ptolemaic theory, for example, accepted the epicyclical loopings of the planets as actually being factual, because they could be clearly observed. Copernicus, in attempting to discover the Pythagorean mathematical order determining the motion of the planets, dismissed these epicyclical motions as mere illusions resulting from the limitations of observations. In this case, a Ptolemaic fact was a Copernican fiction. The raw sensory evidence was virtually identical, but both theories, in their attempt to make the observation comprehensible, arrived at different estimations of its factuality.

The scientist cannot critically analyze his own premises because they establish the principles and standards

---

[28]Polanyi, _Personal Knowledge_, p. 162.

of his thought. Polanyi directs us to a logical circle where "any inquiry into our ultimate beliefs can be consistent only if it presupposes its own conclusions. It must be intentionally circular."[29] He further explains that "there exists no principle of doubt the operation of which will discover for us which of two systems of implicit beliefs is true--except in the sense that we will admit decisive evidence against the one we do not believe to be true, and not against the other."[30] In essence, as William Scott notes, "there is no possibility of a place on which to stand in order to critically doubt the entire basis of any commitment, no way of examining a commitment in a completely non-committal fashion."[31] Thus, even the refuting data scientists might employ to falsify a theory in accordance with Popper's model, requires the constant personal judgment (albeit within the context of the scientific community) of data that is said to be authentically falsifying. Such assessment requires the incorporation of personal standards, beliefs and conceptions of science.

An additional factor that precludes the possibility

---

[29]Polanyi, <u>Personal Knowledge</u>, p. 299.

[30]Ibid., p. 294.

[31]William T. Scott, "Commitment: A Polanyian View," <u>Journal of the British Society for Phenomenology</u> 8, 3 (October 1977):202.

of the scientist critically analyzing the fundamental premises of his understanding of science lies in Polanyi's claim that the premises are functional and non-explicit. They are functional, he explains, in the sense that they are used by the scientist to pursue his studies rather than being themselves objects of science. The scientist is subsidiarily aware of the premises he maintains in order to focus his attention on the objects of science. The premises are not known in and of themselves but only in relation to their function in the processes of science.

The inexplicit character of the premises the scientist accepts is, in essence, a logical consequence of their function; they "can be found authentically manifested [and comprehended] only in the practice of science."[32] The scientist may be able to explicitly state, for example, that he assumes naturalistic premises, but the significance of these premises lies in their shaping of the phenomena the scientist observes. The difference may be illustrated by use of simple analogy.

Both a craftsman and a novice are capable of identifying a particular tool and of indicating some of its functions, but what distinguishes the craftsman's knowledge from that of the novice lies in the former's ability

---

[32]Michael Polanyi, Science, Faith and Society (Chicago and London: The University of Chicago Press, 1946), p. 85.

to use the tool subsidiarily in order to focus on the
object of his craft. In the case of the scientist, he
is not only identified by his explicit knowledge of
given premises, but by his subsidiary utilization of
those premises in the practice of science.[33]

It might be argued that Polanyi has placed us
within an impenetrable circle of irrational and static
belief. This, however, is not the case. Firstly, we
must keep in mind that premises, as they function in
scientific research, are inexplicit and therefore open to
differing functional interpretations. New areas of
inquiry and mechanical and mathematical tools as well as
discoveries require the scientist to apply his functional
knowledge beyond the scope of explicit and objective
rules. This is not to say that specific explicit procedures do not exist but only that they must be applied,
and that the schemata for their application rest upon the
presuppositions the scientist holds. Polanyi states that
"the scientist's procedure is of course methodical. But
his methods are not maxims of an art which he applies in
his own original way to the problem of his choice....
Every factual statement embodies some measure of

---

[33]For example, Lorenz's transformations were known
by physicists for fifty years before Einstein subsidiarily applied them; their implications for physics were
implicit rather than explicit.

responsible judgment as the personal pole of the commitment in which it is affirmed."[34]

Secondly, the premises of science constitute traditions that underly all the processes in science from discovery to evaluation. Polanyi maintains that science is
more than the sum of certain explicit statements referring to and determined by observation. He writes, "to
limit the term science to propositions which we regard as
valid, and the premises of science to what we consider to
be its true premises, is to mutilate our subject matter."[35] Scientific tradition provides the scientist with
a general intellectual orientation (as well as a highly
complex network of explicit theories and procedures)
which circumscribes his rationality and yields implications
for assessment and discovery.

Copernicus's acceptance of Pythagorean metaphysics
facilitated his developing a heliocentric theory of the
universe. His work provided a fruitful set of problems
for Kepler, Galileo, and Newton. It should be noted, however, that Kepler's theory of the eliptical orbit of
Mars, Galileo's treatment of terrestrial phenomena, and
Newton's use of applied mathematics developed out of
problems posed (premises laid by) the Copernican notion

---

[34]Polanyi, Personal Knowledge, pp. 311-312.

[35]Ibid., p. 164.

of the nature of the universe even though they ultimately transformed the premises Copernicus accepted. A
similar transition resulted from Newtonian premises.
Newton "originated the assumption--which was to predomin-
ate until the middle of the nineteenth century--that
science might ultimately reduce all phenomena to the
mechanics of some ultimate constituent particles."[36]
The problems posed by Newtonian premises guided centuries
of scientific inquiry but finally led to their own rejection and an acceptance of physics as a study of "systems
of mathematical invariances."[37] The premises of science
are neither wholly explicit (as they function heuristically) nor static.

The premises that earlier generations of scientists
a-critically assumed are implied in their discoveries
and provide a general foundation for further research.
Polanyi explains that "explicit premises of science are
maxims, which can be acknowledged as such only as part
of a commitment endorsing the scientist's vision of
reality."[38] The acceptance of premises is said to be
a-critical because they are not determined by wholly
rational principles but, rather, provide the forum within which the power of reason can be utilized. Polanyi

---

[36]Polanyi, Science, Faith and Society, pp. 85-86.

[37]Polanyi, Personal Knowledge, p. 164.

[38]Ibid., p. 307.

here calls our attention to the "fiduciary-rootedness of all rationality."[39] The problems faced by all generations of scientists arise within the context of the specific discoveries and general traditions of preceding generations. The finding of rational solutions to scientific problems depends upon the context for inquiry provided by previous research; the discoveries of past generations (inclusive of their general premises) provide the basic clues for discernment of further fruitful fields of inquiry.

Thus, rational scientific propositions are founded upon assumptions "concerning the structure of the universe and on the evidence of observations collected by the method of science...."[40]

Accordingly, new scientific theories incorporate premises assumed in the general scientific community and particular observations and may necessitate a reconsideration of the original premises. The specifics of this process will be presented in greater detail later in this work, but we can now generally state that the implications of given premises can provide problems that can only be coherently solved by rejecting some aspects of the original premises. Kepler, for example, accepted

---

[39]Polanyi, _Personal Knowledge_, p. 164.

[40]Polanyi, _Science, Faith and Society_, p. 31.

Copernican theory but ultimately rejected the idea of the circularity of the orbits of the planets that the theory assumed.

We may conclude that the scientist is not entrapped within state presuppositional scientific traditions, but is given, through them, a context within which to develop new areas of inquiry. In the course of inquiry the scientist, using the inexplicit premises of science, can exercise his own judgment to evaluate experimental evidence and commit himself to new presuppositional frameworks. Such new premises may provide what the scientist believes to be a more suitable framework to exercise his rational powers. The merit of the new orientation will appear as its implications unfold in their application to empirical data.

We may generally say that our subsidiary awareness of the premises we assume form a "heuristic vision" which guides our expectations about what we have yet to observe. It is not being asserted that a given scientist's heuristic vision will have specific predictive value in empirical inquiries, but rather that such vision serves to direct a given scientist toward the specification of problems and choice of potentially productive fields of inquiry. Polanyi concludes that

Tacit assent and intellectual passions [the

desire to develop a rational interpretive
framework], the sharing of an idiom and
of a cultural heritage, affiliation to a
like-minded community: such are the impulses which shape our vision of the nature
of things on which we rely for our mastery
of things.[41]

A heuristic vision is partially shared by all mem-

bers of the scientific community in the general accept-

ance of the premises, theories and methods of present

day science. There is, however, a wholly personal

aspect of the scientist's vision which is dependent upon

his education, intellectual facility, and individual

empirical experience. Without such sociological and

psychological aspects, Polanyi claims that science would

fail to continue its existence. He states:

Our vision of the general nature of things
is our guide for the interpretation of all
future experience. Such guidance is indis-
pensable. Theories of the scientific method
which try to explain the establishment of
scientific truth by any purely objective
formal procedure are doomed to failure. Any
process of inquiry unguided by intellectual
passions would inevitably spread out into a
desert of trivialities.[42]

It may well be argued that Polanyi has merely re-

stated the Humean position that reason has only a minor

role to play in the development of knowledge. Such a

contention, however, fails to recognize that Polanyi

does not share the Humean notion of impersonal

_____

[41]Polanyi, Personal Knowledge, p. 266.

[42]Ibid., p. 135.

objectivity. Rather, he conceives of objectivity in
the metaphysical sense described in the opening sections
of this chapter. It will be recalled that Polanyi does
not believe that the object of science is circumscribed
by empirical observation, but that it is to be found in
the ontological orderliness of nature. Scientific objectivity is thus to be understood as the ideal of rationally representing the laws of nature as they exist independent of and only partially revealed in observation.
We may say, in short, that a theory is objective to the
degree that it provides a rational framework which corresponds to the coherent patterns of nature, or, in
Polanyi's words, "the rationality of nature."

In this case, observed phenomena provide only clues
to orderly patterns in nature or what might be called
rational or coherent. The coherences a scientist
believes to exist ontologically are a result of his
rational integration of empirical observations in accordance with the fiduciary grounds of his own thinking.

Ptolemy and Copernicus, for example, both referred
themselves to the same observations but approached them
with different premises. The resulting coherent entities
they developed (an anthropocentric universe for the former and a heliocentric one for the latter) though both
have proven false, were a result of their fundamental

differences in orientation.[43]   "Copernicus...discovered
new facts because [he] abandoned authority--and not the
other way around."[44]

The coherence a scientist perceives in nature results from his commitment to his own heuristic vision,
and those coherences can not be understood save from
one's commitment to that vision. Since, as noted previously in our discussion of the Tu Quoque question,
Polanyi maintains that all rationality is fiduciarily-
rooted, a scientist cannot stand outside his own commitments to compare another scientist's conclusions with the
"facts"; all a scientist can do is argue against (or for)
another's claim to have discovered a rational coherence
(a natural law) from his own commitments, commitments
that establish and underly the assessment of the rationality of the claim.

For Polanyi, the recognition of ontological order,
order that transcends experience, requires that the scientist believe in a given heuristic vision that will frame
his rational conceptions which will correspond to that
order. "According to the logic of commitment truth is
something that can be thought of only by believing it."[45]

_____

[43]The falsity of the Copernican system here referred
to is relative to Copernicus's insistence upon the per-
fect circularity of planetary orbits among other things.

[44]Polanyi, Science, Faith and Society, p. 26.

[45]Polanyi, Personal Knowledge, p. 305.

It is clear that Polanyi rejects the concept of impersonal objectivity implicit in Hume's quest for justification of inductive inference as well as the positivistic notion that science must remain within the confines of observation. It must also be concluded that he rejects Popper's splitting of Hume's problem into separate psychological and logical questions. The psychological mechanism of belief, for Polanyi, underlies all knowledge and accompanies the scientist in all stages of inquiry from discovery to evaluation. The significance of the scientist's a-critical beliefs which shape his heuristic vision is not limited to discovery; impersonal objectivity cannot, for example, be found in the process of the critical evaluation of scientific theories.

Polanyi explains that

> It is true that a single piece of contradictory evidence refutes a generalization, but experience can present us only with apparent contradictions and there is no strict rule by which to tell whether any apparent contradiction is an actual contradiction. The falsification of a scientific statement can therefore no more be strictly established than can its verification. Verification and falsification are both formally indeterminate procedures.[46]

The case of D.C. Miller clearly illustrates this point. In 1925, Miller presented substantial empirical evidence that was inconsistent with the implication of

------

[46]Michael Polanyi, "The Creative Imagination," Tri-Quarterly 8 (Winter 1967):111.

Einstein's relativity theory. The scientific community, however, was committed to the rational structure that relativity theory provided and rejected and potentially falsifying evidence "in the hope that it would one day turn out to be wrong."[47] The members of the scientific community "had so well closed their minds to any suggestion which threatened the new rationality achieved by Einstein's world-picture, that it was almost impossible for them to think again in different terms."[48]

Apparently contradictory empirical evidence may later prove to be a result of extraneous factors, factors unrecognized at the time of observation. Such appears to have been the case with Miller's results. Had the scientific community not generally committed itself to the rationality of relativity theory, the apparently contradictory data which remained unexplained for many years would have brought about the refutation of a theory we now at the very least find largely unrefuted. Polanyi concludes, "Any critical verification [or falsification] of a scientific statement requires the same powers for recognizing rationality in nature as does the process of scientific discovery even though it exercises these at a lower level."[49]

---

[47]Polanyi, Personal Knowledge, p. 13.

[48]Ibid., p. 13.

[49]Ibid.

The question arises as to how Polanyi unifies his metaphysic with his concept of objectivity. We may focus on the following issue: How can Polanyi avoid subjectivism when the object of science is indeterminate and assailable only from within the context of a system of a-critical beliefs. Polanyi himself notes that such a program threatens to sink into subjectivism: "for by limiting himself to the expression of his own beliefs, the philosopher (or the scientist) may be taken to talk only about himself."[50]

It is necessary to note, in remaining consistent with what we have thus far described, that we will not seek a hard and fast mechanism that will definitively determine the "objective value" of a given theory. Scientists are necessarily committed to sets of a-critical assumptions which structure their visions of reality and underly their determinations about facts and theories; they are involved in a logical circle in which everything must be evaluated in terms of antecedent belief. However disquieting it may be, the eventual justification for any claim of objectivity will necessarily result from the personal responsibility of the scientist and the conviviality of the scientific community.

It is true that the scientist assumes various

---

[50] Ibid., p. 299.

premises within a unique set of conditions, i.e., that
he is born at a distinct place and time, but we need not
assume that the conditions comprise his capacity to be
objective. Polanyi writes, "Believing as I do in the
justification of deliberate intellectual commitments, I
accept these accidents of personal existence as the con-
crete opportunities for exercising our personal responsibility."[51] Although the personal responsibility demanded
of a scientist can only be assumed from within the commitment situation, it enables the scientist to transcend
his or her subjectivity. This apparently logical contra-
diction can be solved by a further definition of the term
"subjectivity."

Polanyi claims that the concept of subjectivity has
been too broad and sweeping. This may well be the result
of a limited concept of objectivity which demands the ex-
trication of all personal factors in science. All personal elements of scientific inquiry, accordingly, have
been assumed to be subjective flaws in the objective body
of science.

Upon closer examination, however, what we generally
call human subjectivity can be broken into two classes
which have distinctly different implications for science.
The first class "is exemplified by dreams, the experience

---

[51]Polanyi, Personal Knowledge, p. 322.

of pain, states of drunken hilarity or pervasive anxiety; states or patterns of behavior, that is, which make no claim external to their subject."[52] The second class consists of those beliefs and standards which we hold to be true and which we believe to have universal validity; in the case of science, the individual submits himself to the concepts and ideas which do not yield to subjective desires or inclinations. It does not matter, with regard to this particular point, whether or not the rationally compelling concepts and ideas are necessarily accurate to the ontological references of inquiry. The critical point is that the individual, in his belief, submits himself to standards which are independent of subjective desire.

It may be said that scientific knowledge is objective, within this limited context, to the extent that it is assumed with _universal intent_, with the belief that the requirements of ontological reality have been met. Polanyi explains, "I speak not of an _established_ universality, but a universal _intent_.... To claim validity for a statement merely declares that it _ought_ to be accepted by all."[53] The scientist's "ought" is not a matter of

---

[52]Brennan, _Journal of the British Society for Phenomenology_, pp. 142-143.

[53]Michael Polanyi, _The Tacit Dimension_ (Garden City: Doubleday & Company, Inc., 1967), p. 78.

caprice but an act of the utmost responsibility.

The subjectivity-objectivity dilemma which has long riddled epistemologists of science can now be conceived as a result of a restrictiveness of the concept of objectivity; in short, it is a result of a lack of recognition of the role of personal responsibility in science. Once it is accepted that scientific knowledge is dependent for its existence on the personal responsibility of the scientist, the logic of the distinctions between subjectivity and objectivity is defused. The personal aspect of a scientist's discovery, evaluation or use of scientific knowledge is neither subjective nor objective as those terms are commonly defined. Polanyi concludes that "in so far as the personal submits to requirements acknowledged by itself, it is not subjective; but in so far as it is an action guided by personal passions, it is not objective either. It transcends the distinction between subjective and objective."[54]

Were this Polanyi's final argument against the threat of subjectivism, his conclusions would prove inadequate to guard against the intrusion of all manners of frauds and incompetents into the scientific arena. Science would run the risk of discontinuity, and ultimately, inertia. Polanyi does, however, offer a

---

[54]Polanyi, Personal Knowledge, p. 300.

consistent response to these objections: the conviviality of science.

Polanyi maintains that through children's a-critical acceptance of the language, custom and modes of thinking of a culture, children are taught the basic cultural premises for interpreting the world. "The naturalistic view held by scientists as by other modern men today has its origin in their primary education."[55] The inculcation of the naturalistic vision of nature is then made more comprehensive and specific in the pursuit of scientific knowledge. "The future scientist is attracted by popular scientific literature or by schoolwork in science long before he can form any true idea of the nature of scientific research."[56] The student acquires a faith in science to reveal mysteries that would otherwise remain hidden. Thus, through his training in science, "his mind will become assimilated to the premises of science. The scientific intuition [vision] of reality henceforth shapes his perception. He learns the methods of scientific investigation and accepts the standards of scientific value."[57]

The development of a scientific attitude begins and progresses with "a recognition of the authority of that which [the individual] is going to learn and of those

---

[55]Polanyi, Science, Faith and Society, p. 42.

[56]Ibid., p. 44.   [57]Ibid.

from whom he is going to learn it."[58] The scientific
community is thus bound together in premises that transcend the particulars of theories. Each member of the
scientific community shares in a convivial pursuit of
scientific truth.

There are, of course, controversies that arise--in-
cluding everything from the design of a given experiment
to the nature of the "truth" being sought--but they are
argued in a common forum and in a common idiom. The decision about the issues are not a matter of majority
rule but of individual conscience. However, the general
acceptability of an argument may very well suffer if the
choices a scientist makes violate, without explanation,
contemporary standards and measures. The threat of discontinuity and fraud is, therefore, minimized by the cultural and scientific traditions of the scientific community.

We have established at this point that Polanyi
believes it is the proper task of science to discover the
orderliness and ontological reality and that the scientific knowledge gained in such an endeavor is neither
strictly verifiable nor falsifiable by observation. We
have also concluded that the scientist, in his judgment,
is held responsible to the demands of ontological reality

---

[58]Polanyi, Science, Faith and Society, p. 45.

as well as to the general standards of the community in which he works. The difficulty which still remains before us, however, is to determine how reality is to be recognized so as to enable the scientist and the scientific community to judge the validity of proposed theories.

It must be emphasized that Polanyi is firmly committed to the use of rigorous experimental techniques and measures, but that the application of techniques and the determination of standards for measures involves the personal discretion of the scientist. He does not deny the need for rational thinking or the importance of empirical evidence, but he does stress that rational thought is fiduciarily-rooted and that empirical evidence is not the sole criterion for objective scientific knowledge. Thus, we may conclude that empirical evidence will play a significant but not wholly dominant role in distinguishing theories responsive to ontological reality from theoretic fictions.

While it is true that factuality is necessarily established within the commitment situation, the sensory content of such determinations can be virtually universal.[59] Copernicus and Ptolemy developed their theories

---

[59]It should be noted here that different presuppositional commitments and even the specifics of a scientist's education may well affect his choice of scientific measures and measurements. With these restrictions, it is

of planetary motion, and established, to their own
satisfaction, the factuality of their claims with reference to virtually identical observations of celestial
activity. The point is that facts are neither capricious
nor insubstantial as a first reading of Polanyi might be
taken to indicate. Empirical evidence provides information for the discovery of structures in nature that can
be rationally interpreted and for the evaluation of scientific claims from within the scientist's presuppositional
position.

The role of empirical evidence can be explained as
follows:

> The part of observation is to supply clues
> for the apprehension of reality: that is,
> the process underlying discovery. The ap-
> prehension of reality thus gained forms in
> its turn a clue to future observations: that
> is the process underlying verification.[60]

This role of empirical information in scientific
discovery and evaluation is logically consistent with
Polanyi's definition of the ontological objectives and
the empirical limitations of the scientist. The orderliness sought is simply not the same as that which is
perceived; perception yields content for rational integration and analysis consistent with the scientist's

---

still legitimate to hold that the sensory evidence in a
given situation is generally publically perceptible.

[60]Polanyi, Science, Faith and Society, p. 29.

fiduciary framework.

The orderliness sought is to be found in the laws operating in an ontological field (a particular area of inquiry). The scientist, in the activities of discovery and evaluation, is subsidiarily aware of empirical evidence, in order to focus his own intention on these laws which are only partially evident. In Polanyi's words, "Scientific knowing consists in discerning gestalten that indicate a true coherence in nature.[61]

It is clear that the scientist has only a vague vision of reality. This is not to say that the theoretic concepts about nature are diffuse but only that they refer to aspects of reality which, in their ontological standing, have an indeterminate range of implications. The distinguishing characteristic of an ontological entity, or a comprehensive entity as it is comprehended by the scientist, lies, according to Polanyi, precisely in its ability to reveal itself in new and varied ways. He states,

> This capacity of a thing to reveal itself
> in unexpected ways in the future I attribute
> to the fact that the thing observed is an
> aspect of reality, possessing a significance
> that is not exhausted by our conception of
> any single aspect of it.[62]

The theory that a scientist applies in his research

---

[61]Polanyi, Science, Faith and Society, p. 10.

[62]Polanyi, The Tacit Dimension, p. 32.

corresponds in its function to a principle (an ontological principle) operating in ontological reality, and individual scientific "facts" are equivalent in their function to empirical observations. If, therefore, a scientific theory accurately corresponds to operations occurring in a comprehensive entity, we can expect that its future implications will accurately reflect the new and varied manifestations of the entity under various conditions. We can conclude, finally, that a theory is objective to the extent that its implications parallel those of comprehensive entities. Polanyi states, "In this wholly indeterminate scope of its true implications lies the deepest sense in which objectivity is attributed to a scientific theory."[63]

It is evident that Polanyi assumes a unique form of verification where the objective validity of a scientific theory is relative to its ability to anticipate and integrate the implications of ontological entities and principles. Scientific theories thus also form part of a heuristic vision guiding the scientitist.

Accordingly, we can see why Polanyi rejects the use of the word "fruitfulness," even Popper's intellectual definition, in the characterization of the objective value of a given theory. The term is relative to that

---

[63]Polanyi, Personal Knowledge, p. 5.

which it refers; it does not stand without qualification.
Polanyi notes that Ptolemaic theory was a _fruitful_ source
of error and astrology a _fruitful_ source of income. A
fruitful scientific theory is one that provides a basis
for its own expansion and conformation in other theories.
Such a characteristic is a result, Polanyi contends, of
the theory's correspondence to or contact with the rational operation of a comprehensive entity. "The Copernican
system did not anticipate the discoveries of Kepler and
Newton accidentally: it led to them _because_ [in essence]
it was true."[64]

The questions arise as to how the scientist can make
a scientific discovery or determine the value of the implications of a theory before he verifies them. The answer to the latter question lies in the scientist's
primary assumptions and the implications of a given
theory. Polanyi states,

> Scientific discovery reveals new knowledge,
> but the new vision which accompanies it is
> not knowledge. It is _less_ than knowledge
> for it is a guess; but it is more than knowl-
> edge, for it is a foreknowledge of things
> yet unknown and at present perhaps inconceiv-
> able.[65]

The process of developing the implications of a scientific
theory and the process of scientific discovery, however,
have yet to be explained.

---

[64]Polanyi, _Personal Knowledge_, p. 147.

[65]Ibid., p. 135.

Chapter II

THE NATURE OF TACIT KNOWING

Thus far we have explored Polanyi's concept of the ontological referents of science. The resultant concept of objectivity was found to incorporate fiduciarily-rooted premises that would hopefully transcend the limitations of present experience to reveal new ontological consistencies in future research. One guard against caprice was found in the personal obligation to work with universal intent within a convivial framework. The present chapter will further define how Polanyi understands the indeterminate objective nature of our scientific knowledge. It will emphasize the role of the presuppositional network which constitutes a heuristic vision of the world and individual areas of research.

Polanyi maintains that scientific inquiry is an effort to create an intellectual framework capable of comprehending ontological reality. He believes that it is only through seeking such an ontological reference that we can use our conceptions to guide our inquiries. He writes,

> Why do we entrust the life and guidance of
> our thoughts to our conceptions? Because
> we believe that their manifest rationality
> is due to their being in contact with

> domains of reality, of which we have
> grasped one aspect. This is why the
> Pygmalion at work in us when we shape
> a conception is ever prepared to seek
> a guidance from his own creation....[1]

Thus, Polanyi assumes a metaphysic which necessarily subsumes the statistical analyses of phenomenal entities--the foundation stone of the logical positivist conception of science--under a fiduciary-rooted[2] attempt to discover the ontological principles underlying the form and structure of that which is perceived. The mechanics of Polanyi's rejection of all formalistic approaches to scientific inquiry will be further refined, as will his general concept of knowing, in this chapter.

In Chapter I we introduced the notion that "scientific knowing consists in discerning gestalten that indicate a true coherence in nature."[3] The term "gestalten" refers to patterns of phenomena. The question arises as to how the scientist can distinguish between a gestalt that indicates an ontological coherence from one that is mathematically consistent yet phenomenally random. A random phenomenal pattern is one that is a function of

---

[1]Michael Polanyi, _Personal Knowledge: Toward a Post-Critical Philosophy_ (New York: Harper & Row, 1964), pp. 5-6.

[2]The fiduciary-rootedness of scientific inquiry is a direct consequence of Polanyi's metaphysic. In so far as reality is said to operate according to laws transcending phenomenal experience they could not be said to be formally derived therefrom. Thus, one's faith in one's underlying intellectual framework provides directions for one's quest for understanding.

[3]Marjorie Grene, _Knowing and Being_ (Berkeley, CA:

the accidental circumstances of observation whereby the
gestalt perceived is the result of factors not under consideration.

We can illustrate the distinction by way of a simple
analogy. If we imagine the objects of experience as
images on a glass, we could clearly describe them and
their relations by an analysis of their spatial configuration. We might attempt to depict the laws governing the
activity of particular entities in accordance with the calculus of their relative relations. Now suppose we step to
one side of the glass to realize that the objects we had
previously understood as two-dimensional are in reality
three-dimensional objects operating in a field behind the
glass. We can then begin to interpret the relations of
the objects from a perspective perpendicular to our original vantage point and perceive a host of interactions
which were hitherto unavailable. Our final conclusion
might be that the original patterns we had perceived were
the incidental results of our position. In short, we might
find that what once had been believed to be "objectively-
derived" knowledge was subjective illusion.

This analogy reflects the historical shift in perspective from Ptolemaic to Copernican theory, and in the more
abstract transition from Newtonian mechanics to relativity

University of California Press, 1974), p. 138.

theory. Just as Copernicus had transcended the empirical particulars of his own situation to discover celestial laws as they would be consistently perceived from non-terrestrial locations, Einstein transcended the limitations of time and space, as they had coexisted in his observation and experience, to conceive of a rational system whereby their "natures" or properties would remain consistent (theoretically consistent) as they underwent mutual transformation in situations beyond those of common experience. There are many schools of thought about the discovery of Copernican theory and Relativity theory, but the fact remains that both achievements incorporated significant conceptual elements that were not available to direct observation or procedurally derivative therefrom. We have previously noted that Copernican theory was contingent upon Copernicus's commitment to a Pythagorean metaphysic not new and Ptolemically inexplicable observation. Further, Karl Popper states, Einstein's relativity theory (as well as the formulations of Einstein's contemporaries) assumed elements that were "highly speculative and abstract, and very far removed from what might be called their 'observational basis.' All attempts to show that they were more or less directly based on observations were unconvincing."[4]

---

[4] Karl Popper, Conjectures and Refutations: The Growth of Scientific Knowledge (New York: Basic Books, Inc., 1965), p. 255.

It must be noted that both Popper and Polanyi agree
that attempts to establish strict procedural guidelines
for the development of scientific theories are misdirected.
Their conceptions of science split as Popper attempts to
construct an impersonal refutationist format (within which
scientific assertions are to be logically assessed), while
Polanyi endeavors to delineate a cognitive model of scientific knowledge that necessitates personal insight to span
the "logical gap" between scientific observation and scientific assertion.

Polanyi's attempt to bridge the gap between observation and scientific assertion with a cognitive or personal
model of knowledge can be clarified as we approach the
question raised before: how to differentiate "random" phenomenal gestalten from phenomenal gestalten indicating
coherent principles operating at an ontological level. To
begin with, the difference between them cannot be discovered or assessed by mathematical procedures. Both types
of gestalten can be mathematically and logically described
in the form of test statements, but their distinguishing
characteristics lie beyond the phenomenal field. This does
not imply that scientists cannot apply formal procedures to
the analysis of data but rather that the final judgment as
to the presence of significance of an ontological principle
in a phenomenal pattern is ultimately informal and personal.
Mathematics can provide formal restrictions on the

evaluation of data, but mathematics alone provides no facility for distinguishing which statistical patterns from among myriads of statistical relations (that can be derived from a given set of data) indicate ontological order. As Poincaré states, "Discovery consists precisely in not constructing useless combinations, but in constructing those that are useful,[5] which are an infinitely small minority."[6]

Judgments as to what fields of study should be pursued, what phenomena should be evaluated, what procedures should be applied in the analysis of data, etc. ... depend secondarily upon predetermined and strictly formal procedures and primarily upon the researcher's personal vision of reality and his commitment to conduct his investigations in accordance with that vision. (It is important to note once again that the underlying presuppositional framework that the scientist assumes is part of a larger cultural vision[7] that has given rise to science and focused its aims. In short, the vision of the scientist is largely culturally and professionally, and with the concept of universal intent, morally shaped and directed. It is not arbitrary or

---

[5]According to Polanyi, the concept of utility here refers to the capacity of a scientific theory to provide a conceptual guide to future contacts with ontological laws.

[6]Henri Poincaré, _Scientific Method_, trans. Francis Martland (New York: Dover Publications, 1952), p. 51.

[7]Further discussions of this issue can be found in E.A. Burtt's _The Metaphysical Foundation of Modern Science_

haphazard.)

Mathematics, for example, plays a highly significant role in the process of scientific inquiry, but it is a subordinate role. Polanyi explains,

> Mathematics only inserts a formalized link in a procedure which starts with the intuitive surmise of a significant shape, and ends with an equally informal decision to reject or accept it as truly significant by considering the computed numerical probability of its being accidental.[8]

The term "intuitive" can be defined for our present purposes as a personal process of intellectual judgment that cannot be formally explicated.[9] The meaning of the term will become increasingly clear through our examination of Polanyi's model of scientific knowing as well as the subsequent discussion of his notion of scientific discovery. (The concept of "intuition" will be fully developed in Chapters III and IV.) He clarifies the superordinate role of the intuitive and informal aspects of knowledge by stating:

> ...we can use our formulas only after we have made sense of the world to the point of asking questions about it and have established the bearing of the formulas on the experience that they are to explain. Mathematical reasoning about experience must include, besides the antecedent non-mathematical finding and shaping of the experience, the equally non-mathematical

and some of the works of Joseph Agassi; of particular interest are Agassi's "Dinosaurs and Horses, or: Ways with Nature," Synthesis 32 (Nov.-Dec. 1975):233-247; "Rationality and the 'Tu Quoque' Argument," Inquiry 16 (Winter 1973):395-406; and "Sociologism in Philosophy of Science," Metaphilosophy 3 (April 1972):103-122.

[8]Grene, Knowing and Being, p. 132.

[9]Ibid., p. 179.

> relating of mathematics to such experience,
> and the eventual, also non-mathematical,
> understanding of experience elucidated by
> mathematical theory. It must also include
> ourselves, carrying out and committing our-
> selves, to these non-mathematical acts of
> knowing.[10]

Polanyi does not here imply that all scientists must individually develop procedural guidelines, but rather, he intends to call attention to the fact that such rules are based, whether historically or person-ally, on informal assessments of what would constitute valid assertions and acceptable standards and measures. Thus, we must conclude that the discernment of gestalten indicating ontological laws cannot be totally accomp-lished by the use of impersonal formal or mathematical procedures but requires personal assessments and personal vision. No analysis of the relative relations of observed entities could distinguish, apodictically, a random series of observations from one indicating an on-tological law.

The process of discerning gestalten that indicate coherent ontological structures or principles and the process of assessing the capacity of a given theory to rationally portray such laws of nature are guided through-out by the personal sense of intellectual satisfaction based upon the scientist's fiduciarily-rooted vision of

---

[10]Grene, _Knowing and Being_, p. 179.

reality. Consequently, scientific knowledge necessarily
includes an element of personal participation. The degree
of personal involvement decreases as observations and
assertions become increasingly routine; where there is
little doubt, there is little impetus for an evaluation
of one's fundamental beliefs. Polanyi writes, "...no
sincere assertion of facts is essentially unaccompanied
by feelings of intellectual satisfaction or of a persua-
sive desire and a sense of personal responsibility."[11]
He further explains,

> Unless an assertion of fact is accompanied
> by some heuristic or pervasive feeling, it
> is a mere form of words saying nothing. Any
> attempt to eliminate this personal co-effi-
> cient by laying down precise rules for mak-
> ing or testing assertions of fact, is con-
> demned to futility from the start.[12]

Thus, the scientific theories that fill the pages of
scientific texts and journals cannot, in themselves, be
classified as cases of scientific knowledge. Each scientific assertion carries with it an element of personal
commitment and responsibility; each scientific assertion
is tacitly accredited. Without their a-critical and
personal assimilation by individuals, scientific theories
are but abstract marks on a sheet of paper. It is only
when the theory is understood to point beyond itself, to

---

[11]Polanyi, _Personal Knowledge_, p. 271.

[12]Ibid., p. 254.

to make a particular aspect of reality rationally comprehensible, that it may be said to have meaning.

Polanyi maintains that scientific theories employ mathematical procedures to provide formal conceptual structures in accordance with an underlying vision of reality. As such, scientific theories can be considered the formalized efforts of one's presuppositional vision. Scientific theories are like tools or probes that more or less satisfactorily meet the intellectual requirements of the scientist's primary intellectual tenets. Further, scientific theories are interpretive frameworks that can only be fully understood as they operate rather than as fixed and isolated objects. To have knowledge of a theory is to be able to apply it to new and unique situations. Thus, Polanyi concludes,

> Knowledge is an activity which would be better described as a process of knowing. Indeed, as the scientist goes on inquiring into yet uncomprehended experiences, so do those who accept his discoveries as established knowledge keep applying this to ever changing situations, developing it each time a step further. Research is an intensely dynamic inquiring, while knowledge is a more quiet research.[13]

The process of knowing does not end with the acquisition of the mechanics of a given theory but includes the assimilation of the vision structuring that theory. A scientific theory both provides an explicit means of

---

[13]Grene, _Knowing and Being_, p. 132.

interpreting a given field of study and a presuppositional orientation. The latter contains implications
for new areas of research as well as new approaches to
fields already under investigation. The problem a given
theory addresses, the methods it employs, the language
it incorporates, the factors it focuses upon, etc. ...
are grounded in a heuristic vision. The assimilation of
that heuristic vision serves to guide the researcher in
his pursuits. To employ a scientific theory as an interpretive framework is to assume the general concepts underlying its capacity to make reality comprehensible. In so
far as one does not assimilate the tacit foundations of
a scientific theory, the performance of the procedures
prescribed by the theory is a purely deductive exercise
having no ontological reference. The establishment of
coherence requires an informal preconception of acceptable
order and regularity.

This is not to say that a scientist must accept the
standards and implications underlying a given theory in
order to assess it. What is being said is that a theory
will be as useful or as meaningful to a scientist as it is
in responding to the requirements of his heuristic vision.
To maintain that a scientist can function in a purely
deductive framework without heuristic underpinnings is
philosophically and practically problematic.

Philosophically, as has been stated, the scientist

can only accept or reject a given theory on the basis of presuppositions and a-critically-grounded formal structures. Surprisingly, we can find that even Karl Popper recognizes the role of a substantial background against which scientific discussion can be carried out. He states,

> People involved in a fruitful critical discussion of a problem often rely if only unconsciously upon two things: the acceptance by all parties of the common aim of getting at the truth, and a considerable amount of common background knowledge....[14,15]

With regard to the practical aspects of the issue, Polanyi states,

> We can derive rules of observation and verification only from examples of factual statements that we have accepted as true <u>before</u> we knew the rules; and <u>in the end</u> the application of our rules will necessarily fall back once more on factual observations, the acceptance of which is an act of personal judgment, unguided by any explicit rules. And besides, the application of such rules must rely all the time on the guidance of our personal judgment.[16]

Yet further, he explains that,

---

[14] Popper, <u>Conjectures and Refutations</u>, p. 238.

[15] The central difference between Popper and Polanyi with regard to this issue lies in the fact that Popper believes that all of our background knowledge is explicated in a piecemeal fashion while Polanyi maintains that common knowledge is partially presuppositional and informally active in the creation of explicit concepts.

[16] Polanyi, <u>Personal Knowledge</u>, p. 254.

> ...every step made in the pursuit of
> science is <u>definitive</u>, definitive in
> the vital sense that it definitely dis-
> poses of the time, the effort, and the
> material resources used in making that
> step. Such investments add up with
> frightening speed to the whole profes-
> sional life of the scientist.[17]

Consequently, the theories that a scientist decides to

investigate or to apply as a basis for future research

are carefully selected so as to bear a maximum benefit

from the efforts required. The problem addressed, the

method of approach, and other significant factors incor-

porated in the development of a scientific theory, all

depend upon a pioneering scientist's commitment to an

image of reality. Similarly, the initial estimations of

the value of the theory by other scientists include their

sense of intellectual satisfaction or lack thereof with

the presuppositions underlying the theory (as well as the

empirical accuracy of a given theory). Just as the dis-

cernment of significant statistical patterns has been

shown to be a fiduciarily-rooted endeavor, so we can now

recognize that the subsequent attention given to a scien-

tific theory is relative to the intellectual satisfaction

it bears to the heuristic premises of other scientists.

Certainly, logical arguments and formal procedures are

involved in the evaluation of scientific theories, but

---

[17]Michael Polanyi, "Logic and Psychology," <u>American Psychologist</u> 23, 2 (January 1968):41.

such formal structures are themselves only as pertinent and as significant as scientists accredit them to be--a judgment, again, tacitly rooted.

Thus, Polanyi reaffirms that "nothing that is said, written or printed can ever mean anything in itself; for it is only a person who writes something--or who listens to it or reads it--who can mean something by it."[18]

A scientific theory, in as far as it can be said to constitute an instance of scientific knowledge, therefore contains not only a logical structure but one that is cognitive and dependent upon a personal knower.

The cognitive aspect of knowledge is unspecifiable in the sense that its explication would dissolve its function. Polanyi explains that this personal and tacit dimension of scientific knowledge is similar to the individual skill that a master craftsman applies in the performance of his or her task. He states, "Though we may prefer to speak of understanding a comprehensive entity or situation and of mastering a skill, we do use the words nearly as synonyms. Actually, we speak equally of grasping a subject or an art."[19] The similarity of the tacit cognitive element of a scientist's knowledge and a craftsman's skill lies in the fact that both activities

---

[18]Michael Polanyi, Study of Man (Chicago, IL: The University of Chicago Press, 1958), p. 22.

[19]Polanyi, Personal Knowledge, p. XIII.

require the individual to exercise personal judgment, analytically irreducible judgment, in the utilization of a given entity (a tool or scientific theory).

The craftsman, while performing his art, considers his tools as they relate to the objects of his activity. The painter, for example, attends to his brush not as a wooden implement but as it functions within the context of his work; he attends to the colors on his pallet not as fluids reflecting various frequencies of light but as hues more or less successful in meeting his needs. Certainly, brushes and paints can be explicitly described, and they may even be identified as having specific functions. However, no amount of analysis can define them as they are functionally apprehended by the artist. In the process of using these items their identification shifts from an emphasis on their isolated characteristics to an emphasis on their operational properties within the context of the artist's intentions. The artist's knowledge of his materials may be said to be inexplicit in the sense that their properties cannot be fully defined by an analysis of their physical characteristics or the calculus of their movement.

Polanyi maintains that a scientist's knowledge of a scientific theory shares similar characteristics. The scientist's knowledge of a given theory is equivalent to the artist's knowledge of his tools and materials. In both cases, the knower can be said to have knowledge to the extent that he can apply that which is known to a given field which constitutes its object. The artist applies his knowledge of his materials to the creation of a coherent image; the scientist applies his knowledge of scientific theory to integrate phenomena into a rational coherence. The major difference between the two skillful performances lies in the responsibility of the scientist to act, not in accordance with subjective inclination but rather in accordance with his universal intention.

The opening sections of this chapter have presented Polanyi's contention that the discovery, implementation and assessment of scientific theories are fiduciarily-rooted processes within which formal and mathematical procedures play ancillary roles. The value of a scientific theory has been understood to be dependent upon its ability to meet the intellectual requirements of a scientist's presuppositional heuristic vision of reality and corollary the given field of study. We must now add that the ability to integrate data into coherent entities or into scientific problems that might reveal future coherences is similarly a personal act of skillful integration. Just as the artist uses his materials to convey or depict an underlying theme or image, the scientist applies scientific theories to observed particulars to discern coherent entities that his heuristic vision of reality would guide him to anticipate. The guidance here is not only in the form of specific content and predictions; it includes a general orientation, an approach, an indication of areas where new and fruitful problems can be pursued.

Scientific knowledge requires a fundamental context that renders explicit assertions functional. Scientific theories may be said to be functional to the extent that they provide rational interpretive systems that are incorporated by the heuristic presuppositions of the scientist. The relative success or failure of a given theory and its implications for new areas of research, therefore, require that a scientist not merely have a critical grasp of the theory's content but also a personal conception of its orientation and intent. The sheer management of phenomena is an insufficient indicator to determine the future value of a theory as there is no formal procedure to determine which, from among an infinite variety of statistical relations in the phenomenal field, are indicative of ontological order and hence future confirmations. The assessment and use of a given theory depend upon the scientist's presuppositionally-grounded sense of intellectual satisfaction with the rationality implicit in the framing of the theory as well as his critical analysis of the technical structure of the formal theory itself.

The knower is unspecifiably and personally active in the process of knowing even though systematic and formal procedures may be used. Polanyi writes, knowing "is a performance, like understanding or meaning something, which can be done only in our heads and not by operating with signs on paper."[20]

We can illustrate Polanyi's conclusion in the following way. Let us assume for the moment that we are highly scientific travelers from another planet who have stumbled upon a wrist watch. We might begin our scientific analysis of the object by weighing it, measuring it, and estimating its volume. A second series of procedures might include analyzing the number and ratios of its gears and then analyzing their metallic composition. Upon completing our analysis, we might conclude that we have a thorough knowledge of the watch and its workings. Such analyses, however, would never reveal that the watch has a function or whether or not the watch in its present condition is capable of successfully fulfilling that function. The purpose of the watch, the object toward which its individual parts were directed, could not be found through an analysis of the parts themselves.

Similarly, the logical analysis of the content of a scientific theory would not indicate its purpose or provide for the determination of its relative success or failure. The presuppositional foundations of the theory--

---

[20]Polanyi, Study of Man, p. 25.

the tacit beliefs that underly its claims to ration-
ality--must be assumed in the process of knowing it.
The argument that the progress of science and the meaning
of a scientific theory can be fully explicated and formalized fails to take full account of the contention that
presuppositions undergird rational inquiry and all language. In addition to the fact that such presuppositions
cannot be circumscribed by formal structures, they are
heuristic and can only be fully understood as they serve
in that capacity. Polanyi quite generally concludes that

> Words can convey information, a series of
> algebraic symbols can constitute a mathe-
> matical deduction, a map can set out the
> topography of a region; but neither words
> nor symbols nor maps can be said to com-
> municate an understanding of themselves.
> Though such statements will be made in a
> form which best induces an understanding of
> their message, the sender of the message
> will always have to rely for the comprehen-
> sion of his message on the [informal] in-
> telligence of the person addressed.[21]

Consequently, knowing a scientific theory implies
both that one is critically aware of its formal structure
and tacitly aware of its unspecifiable presuppositions
and implications. The second element is not formal or
analytic or capable of such depiction; it is psychological
rather than logical. Polanyi states, "Science is grounded,
and firmly grounded, on the indefinable insight which the
current view of science regards as mere psychological

---

[21]Polanyi, Study of Man, pp. 21-22.

phenomena, incapable of producing rational inferences."[22]
Knowing is a process whose mechanisms require an informal
integration of phenomena into coherent entities.

We can now see in the knower a structure similar to
that which is known. Chapter I presented Polanyi's contention that phenomena are governed by ontological laws
or operational principles that transcend the particulars
of the phenomenal field being considered. Corollary, we
concluded that as such operational principles extend beyond
the parameters of experience, they can be expected to reveal themselves in new and unforeseen ways. The statistical relations between phenomena and the characteristics of
the phenomena themselves are consistent with such operations' principles but cannot portray them definitively.
In the case of the knower, the individual's vision transcends and provides for the development and assessment of
scientific theories. Thus, the presuppositional notion
of reality fundamental to the scientist's understanding
of a theory is rightly called a heuristic vision; its
implications extend beyond the parameters of explicit
formulations. Polanyi writes that it thus "seems plaus-
ible to assume in all...instances of tacit knowing the
correspondence between the structure of comprehension and
the structure of the comprehensive entity which is its

---

[22]Polanyi, American Psychologist, p. 27.

object."[23] The formal structures constituting all
theories are consistent with the implications of the
vision guiding the scientist but are incapable of representing it in full.

This structural correspondence between the knower and
the known provides a bridge between Polanyi's metaphysic
and his heuristic epistemology. The operational principles active in reality are not static statistical relations but dynamic characteristics, latent potentialities,
that emerge in different ways depending upon the specific
conditions of the situation in question. Under a variety
of conditions, an operational principle will manifest itself in a variety of ways. Just as we assume an individual will react in new and varied ways, Polanyi suggests
that we can expect that an actual operational principle
will, under different circumstances, bear new but fundamentally consistent implications. Thus, operational principles cannot be circumscribed by observation and their
implications cannot be derived from analyses of data.
Operational principles are phenomenally indeterminate and
formally unspecifiable. Consequently, scientific theories
have an indeterminate and formally unspecifiable aspect to
their value.

The heuristic vision underlying a given scientist's

---

[23]Michael Polanyi, _The Tacit Dimension_ (Garden City,
NY: Doubleday & Company, 1966), pp. 33-34.

theoretical understanding of reality is a presuppositional matrix that provides the foundation of his rationality.
It cannot be reduced to specific rational formulae but is
itself responsible for the determination of the rational
status of explicit statements. A scientist understands
his heuristic vision as it actively functions in the integration of phenomena into rationally consistent conceptual
entities. Scientific theories solidify and articulate the
results of that process (to be described in greater detail
in Chapters III and IV), and science is a unified effort
in so far as scientific theories in numerous fields are
consistent with a convivial vision (a mutually accepted
presuppositional matrix) that renders all such theories
rational. Just as operational principles provide a unified and consistent yet phenomenally indeterminate basis
for observable events, the scientist's heuristic vision--
which he shares in part with all other members of the scientific community--provides a unified and consistent yet
logically irreducible basis for the development of rational
scientific theories. When Polanyi states that the intention of scientific inquiry is to discover the rationality
of nature, he expresses the underlying assumption that
the heuristic vision guiding scientific inquiry can
establish a framework for rationality that corresponds in
its implications to the implications of the operational
principles functioning in the ontological domain.

We have thus far analyzed the tenets of Polanyi's arguments for a heuristic epistemology in science, but the question still remains as to how the tacit foundations of a scientist's intellectual framework operate. It is essential to bear in mind that the tacit dimension of science is not explicit or capable of logical exposition; it is fundamentally psychological. Polanyi comments in this regard,

> My main task will be to survey the non-strict rules of inference—in other words, the informal logic—on which science rests. This non-strict logic will be seen to rest to some extent on psychological observations not hitherto accepted as the foundation of scientific inference.[24]

We noted in Chapter I that Polanyi's metaphysic necessarily implied that phenomena be understood as clues indicating ontological coherences rather than as the exclusive content or final measure of science. It is now clear that the integration of phenomenal clues is not a logical process. Polanyi maintains that the informal mechanisms employed in the process of perception are paradigmatic for the process of scientific inquiry. Polanyi explains,

> While the integration of clues to perceptions may be virtually effortless, the integration of clues to discoveries may require sustained efforts guided by exceptional gifts. But the difference is only

---

[24]Polanyi, _American Psychologist_, p. 27.

one of range and degree: the transition
from perception to discovery is unbroken.
The logic of perceptual integration may
serve therefore as a model for the logic
of discovery.[25]

Polanyi focuses his analysis of the perceptual pro-

cess upon experimental evidence gathered by Gestalt

psychologists. The correspondence of the Gestalt ap-

proach to Polanyi's own orientation lies in the fact that

Gestalt psychologists operate under the notion that per-

ceived particulars are not seen in isolation but as parts

of a configuration of phenomena. Similarly, Polanyi con-

ceives of scientific knowing as an integration of phenom-

ena into rational coherences. The major differences in

the two positions lies in the fact that "Gestalt psychol-

ogy has assumed that perception of a physiognomy[26] takes

place through the spontaneous equilibrium of its particu-

lars impressed on the retina or on the brain."[27] Polanyi,

on the contrary, maintains that the integrative process

(both in the act of perception and in the conduct of

inquiry) is an "active shaping of experience performed in

the pursuit of knowledge."[28] He understands the use of

---

[25]Grene, Knowing and Being, p. 139.

[26]Polanyi uses the term "physiognomy" quite generally to denote a configuration of phenomena that constitute
the outline of a comprehensive entity. The symptomatology
of a disease, for example, is said to be its physiognomy.

[27]Polanyi, The Tacit Dimension, p. 6.

[28]Ibid.

phenomena as clues indicating coherences not as physio-
logically fixed but heuristically framed by the perceiving subject. This responsible integration allows preconceptions and explicit conceptions to affect the perceptual process and ultimately provides a bridge to the
complex integration of phenomena in scientific inquiry.
Polanyi maintains that scientific inquiry differs from
ordinary perception only in "the fact that it can integrate shapes that ordinary perception cannot readily
handle."[29]

A paradigmatic case illustrating the operational
principles of the process of scientific knowing is found
in a Gestalt psychology experiment by Lazarus and Mc-
Cleary (1949). In this experiment the researchers presented subjects with syllables and followed some of the
presentations with a mild electric shock. Eventually,
subjects began to anticipate the shocks with the presentation of certain syllables. They were unable, however,
to specifically identify those syllables that were associated with the shocks. Polanyi notes that the subjects
"had come to know when to expect a shock, but...could not
tell what made [them] expect it."[30] For the sake of clarity, we can refer to the syllables as the first term and

---

[29]Grene, Knowing and Being, p. 138.

[30]Polanyi, The Tacit Dimension, p. 8.

the shocks as the second term of knowing.

The first term was not perceived in isolation; it did not constitute a distinct entity to be considered as an object of contemplation or study. Rather, the first term was perceived within the context of its association with the second term. That is, it was important to the subjects only in so far as it was related to, or capable of providing information about, the second term. Thus, the first term was identifiable not as a distinct object but rather as an instrument or tool that could facilitate the subject's ability to attend to the second term.

The second term was specifically identifiable as it occupied the focus of the subject's attention. The subjects not only were able to explicitly identify the second term but also to transpose other elements constituting the first term into descriptors or indicators of its activity.

The critical factor distinguishing the two types of knowledge lies not in their content but in their psychological function. The items constituting the first term are known subsidiarily. They function as clues to enable the knower to form a consistent perceptual or conceptual image of the object being considered. We may note, for example, that we commonly recognize an individual by the features of his face even though we may not often be able

to explicitly state his particular distinguishing physiognomatic characteristics. Polanyi provides a more
clinical example in the case of a noted psychiatrist
and a group of his students who observed a patient undergo some kind of mild fit. In the discussion that ensued,
the issue was raised as to the specific definition of the
seizure. The discussion concluded with the psychiatrist
stating, "Gentlemen...you have seen a true epileptic
seizure. I cannot tell you how to recognize it; you will
learn this by more extensive experience."[31]

This example illustrates that perceptual particulars can and do play a subsidiary role in the recognition
of a perceptual object. Polanyi writes,

> Whenever we are focusing our attention on a
> particular object, we are relying for doing
> so on our awareness of many things to which
> we are not attending directly at the moment,
> but which are yet functioning as compelling
> clues for the way the object of our attention
> will appear to our senses.[32]

The items constituting the second term are not
known subsidiarily but rather focally. When we recognize
an individual, we may be said to be subsidiarily aware
of the features of his or her face and to be focally
aware of their joint meaning. In the case of the
psychiatrist, he could be said to be subsidiarily aware

[31]Grene, Knowing and Being, p. 125.

[32]Ibid., p. 113.

of the particulars of the patient's symptoms and to be focally aware of their meaning as a single comprehensive entity, i.e., a given physiological or psychological disorder.

It might be objected that in presenting the notions of subsidiary and focal awareness we have altered the character of the original laboratory experiment. In that case, the first term and the second term of the knowing process were not given within the same perceptual field; they were fundamentally related in time. In the physiognomatic example, the first and second terms were enclosed within a given perceptual field; they were fundamentally related in space. However, such distinctions are peripheral to the present philosophical analysis of the structure of the activity of perception, and ultimately, scientific inquiry: the subsidiary/focal relation of objects operates in both space and time.

The significance of the subsidiary/focal distinction lies in its identification of an element of all human knowing that is, but its nature, nonspecific and indeterminate. The non-specific character of the subsidiary elements of the environment, whether they be spatially or temporally related to a given focal object form what Gestalt psychologists call a background. It is only by virtue of this background that the focal object can be identified as such.

The subsidiary elements that enable an individual
to perceive an object or to conceive of its nature are
highly varied. In a sample case of perception, for example, Polanyi contends that we are subsidiarily aware
of the bearing of the fine eye adjustments we make in
perceiving an object clearly. "It is the supreme achievement of our eyes to show us objects as having a constant
colour, size and shape, irrespective of their distance,
position and illumination."[33] This is not at all to say
that such triumphs are a function of the eyes alone but
only to note that they play a necessary subsidiary role, a
role of which we are at best dimly aware, in enabling us
to focus ourselves on a given perceptual object.

We have already noted that subsidiary roles can be
played by objects found in a perceptual field (as in the
Lazarus and McCleary experiment). It must here be stated
that subsidiary information can also be conceptual as in
the case of the psychiatrist diagnosing the seizure. The
diagnosis, the identification of a specific disorder, required a massive theoretical as well as experiental background. It was indeed necessary for the medical students
to acquire personal experience if they were to learn to
make an accurate diagnosis, but this ability to integrate
the subsidiary clues into a unified, consistent and

---

[33]Grene, Knowing and Being, p. 114.

comprehensive framework, into a clear focus, was necessarily dependent upon their subsidiary knowledge of
psychiatric theory.

It might be argued that while Polanyi's analyses
may have validity in cases involving perception or practical skills, it is spurious where abstract and intellectual processes are being considered. The argument
may be presented in the following manner. The unspecifiability of the former category of subsidiary awareness is
a function of the necessity to involve the manipulation
of physical entities. The operations performed thus lack
universally applicable standards and are usually assessed
in accordance with their functional success or personal
preference. In scientific matters, however, the perceptual
and conceptual information that may be said to occupy the
first position can be clearly and explicitly identified.
The tacit quality, the unspecifiability character of the
subsidiary element of the knowing process could, therefore, be eliminated.

This objection, however, fails to take full account
of Polanyi's subsidiary/focal distinction. As we noted
earlier, the essential difference between the subsidiary
and focal elements of the knowing process is psychological rather than logical. What is essential is not the
content of the two elements but the function of the
content in the mind of the knower.

Polanyi explains,

> The essential feature throughout (all
> instances of knowing) is the fact that
> <u>particulars can be noticed in two dif-
> ferent ways</u>. We can be aware of them
> uncomprehendingly, i.e., in themselves,
> or understandingly, in their participa-
> tion in a comprehensive entity. In the
> first case we focus our attention on the
> isolated particulars; in the second, our
> attention is directed beyond them to the
> entity to which they contribute....
>
> ...Focal and subsidiary awareness are
> definitely <u>not two degrees</u> of attention
> but <u>two kinds</u> of attention given to the
> same particulars.[34]

Thus, subsidiary entities are not merely objects

that have remained for any number of reasons unformal-

ized, but they are, by virtue of their psychological po-

sition, unformalizable. It may be that, if given suffi-

cient time, we might be able to identify the specific

qualities of a familiar face, and given a similar oppor-

tunity, the distinguished psychiatrist could identify

the publicly observable aspects of a patient's behavior

that form the basis of his diagnosis. In assuming such

a task, however, the individual is called upon to shift

his own attention from a subsidiary awareness of particu-

lar phenomena (from the phenomena as they bear upon a

focal object) to a focal awareness of the phenomena them-

selves. Whereas the individual had once referred <u>from</u>

---

[34]Ibid., p. 128.

subsidiary phenomena <u>to</u> a focal object, he would now be referring <u>to</u> the previously subsidiary phenomena <u>from</u> other particulars. In a word, in reorganizing one's focus, one has transformed the psychological position and character of the objects which constitute one's reference.

The specific transformation of the character of an object when it is shifted from a subsidiary to a focal entity can be further clarified through an analysis of what Polyani calls "the triad of tacit knowing." Polanyi explains that "the triad of tacit knowing consists in subsidiary things (B) bearing on a focus (C) by virtue of an integration performed by a person (A)..."[35] A significant element for our present purpose is the person who performs the integration. The bearing of B on C is relative to the activity of the knower and this "from-to relation lasts only so long as a person, the knower, sustains this integration."[36] Detaching B from its subsidiary role in A's focusing on C would dissolve its psychological character and <u>dis</u>-integrate the triadic structure of tacit knowing. The detached definition of B as a focal object would serve to transform it into a new object in accordance with its new focal

---

[35]Grene, <u>Knowing and Being</u>, p. 182.

[36]Polanyi, <u>American Psychologist</u>, p. 31.

emphasis. B is, therefore, logically unspecifiable not as an object, but as it functions in the knowing process. Generally stated, "...our attention can hold only one focus as a time and...it would hence be self-contradictory to be both subsidiary and focally aware of the same particulars at the same time.[37] All that which acts subsidiarily is endowed with a vectorial quality that eludes inspection when an object is considered in and of itself.

Examples of this logical unspecifiability can be found in virtually all areas of life requiring some kind of skillful performance. A blind man using a walking stick interprets the vibrations he feels in his palm as they inform him of his surroundings. He does not attend to the vibrations themselves, to the feel of the stick as it rests in his hand. Such a change in focus would render his walking stick virtually useless. A pianist playing a concert would paralyze his own performance were he to shift his attention from his music to his fingers pressing against the keyboard. The concert goer would fail to hear the music were he to focus upon the melodies and harmonies as intervals of sounds of varying frequencies and amplitudes.

Similarly, a logical unspecifiability pervades the process of scientific knowledge. Every equation and every

---

[37]Polanyi, Personal Knowledge, p. 57.

assertion directs the scientist toward a comprehensive entity, toward a rational conception of the principle operating in reality. Although the process of knowing will eventually reveal itself to be far more complex than it may now appear, we can now simply state that as a scientific theory is used to focus research problems or to interpret data, it occupies a subsidiary position in the tacit knowing triad.

It is now necessary--if we are to develop a more comprehensive and consistent concept of the process of scientific knowledge--that we first clarify the relationship between Polanyi's subsidiary/focal distinction and the general notion of consciousness.* The subsidiary/focal distinction is a fundamental characteristic of consciousness. That which is subsidiary is not equivalent to that which is subconscious or unconscious; "it can range from a subliminal level to a fully conscious level."[38] The scientist's application of a given scientific theory proceeds with a full recognition of his efforts. Indeed, Polyani maintains that the scientist's judgment is at all times based upon a subsidiary awareness of his heuristic vision of reality, but this is not to say that the scientist is necessarily unconscious or subconsciously doing so. To

---

[38]Polanyi, _American Psychologist_, p. 31.

*Polanyi's concept of consciousness will be more fully explored in Chapter III.

be conscious of something does not necessarily mean that the object is specifiable. The unspecifiable subsidiary element in the structure of knowing is a necessary component in the diffuse or crystalline awareness of an entity. Polanyi summarizes this idea by stating that, "what makes an awareness subsidiary is the function it fulfills; it can have any degree of consciousness, so long as it functions as a clue to the object of our focal attention."[39]

The preceding argument raises the question as to the relation between Polanyi's subsidiary/focal and tacit/explicit distinctions. It may have first appeared that these distinctions were equivalent, but it now appears that subsidiary factors play a significant role in the attainment and comprehension of explicit as well as tacit knowledge.

Although the first and second terms of the tacit knowing triad have been viewed as parts of a single knowing process, Polanyi maintains that they represent two distinct but interrelated kinds of knowledge. The specifiability of the second term illustrates the key characteristic of what Polanyi calls explicit knowledge. Such knowledge is formal and logically structured; it is capable of being critically analyzed and reviewed. Maps, mathematical formulae and the formal structures of scientific theories fall into this classification. The

---

[39]Polanyi, The Tacit Dimension, pp. 95-96.

unspecifiability of the first term illustrates the most
notable characteristic of what Polanyi calls <u>tacit</u> knowledge. Tacit knowledge is informal and dynamic, personal
and purposive; it is an unformalizable guide in judgment
in matters requiring the synthesis or integration of
phenomena whether physical (as in the case of mastering a
craft) or conceptual (as in the case of discovering a scientific theory). The drawing of maps, the formulating of new
geometries, and Einstein's envisioning of the universe from
the leading edge of a beam of light rely upon tacit knowledge. Polanyi maintains that tacit and explicit knowledge
are not separate or mutually exclusive.

The difference between tacit and explicit knowledge
may be explained in terms of an object's position in the
tacit knowing triad. An individual may have knowledge of
an entity (B) as it functions subsidiarily to focus upon
something else (C). As stated previously, although the
individual might well be able to clearly identify the subsidiarily functioning object (B), he can do so only by dis-
integrating its vectorial character in the tacit knowing
triad. We may thus conclude that such knowledge of B is
necessarily unspecifiable and therefore tacit.

Contrastingly, an individual may have knowledge of
an entity (C) as it is brought into focus by numerous subsidiary factors (B); an individual may, in short, be able
to clearly and distinctly identify a given focal entity.

In this case, the individual may be said to have an explicit knowledge of that entity (C). Such knowledge, however, implies that the individual has integrated numerous subsidiary clues in order to yield that focus. Without the subsidiary knowledge of numerous entities (B) the focal object (C) could not be attended to or specified. Explicit knowledge may, therefore, be said to be tacitly rooted. "While tacit knowledge can be possessed by itself, explicit knowledge must rely on being tacitly understood and applied. Hence, all knowledge is either tacit or rooted in tacit knowledge."[40]

Thus it may be said that the focal/subsidiary distinction transcends the explicit/tacit distinction. Explicit knowledge always necessitates that the knower bring subsidiaries to bear on a focus; all explicit knowledge has both a subsidiary and a focal aspect.[41] Polanyi concludes,

> The ideal of a strictly explicit knowledge
> is indeed self-contradictory; deprived of
> their tacit co-efficients, all spoken
> words, all formulae, all maps and graphs are
> strictly meaningless. An exact mathemati-
> cal theory means nothing unless we recognize
> an inexact non-mathematical knowledge on
> which it bears and a person whose judgment

---

[40]Polanyi, The Tacit Dimension, pp. 95-96.

[41]Tacit knowledge implies that one can integrate subsidiaries to bear upon a focus. Tacit knowledge therefore operates when a focal object is introduced.

upholds this bearing.[42,43]

We have previously stated that knowledge is an activity that requires skill. We can now state, more specifically, that knowing is an integrative process whose tacit elements govern the formulation and use of all formal procedures and explicit scientific theories.[44] Polanyi refers to the integrated process as an "act of tacit inference."

Formal procedures and theories play a highly significant role in the process of scientific inquiry, but decisions as to the proper application of various procedures and theories and judgments regarding their import require that the scientist perform informal tacit appraisals of their utility. The formalisms of science, when they are applied in scientific inquiry, become subsidiary factors in the scientist's cognitive framework, and they are applied in accordance with the scientist's skill in integrating them with data to reveal significant patters and order.

It may be argued that the formalisms of science have become so refined that the personal judgment of the scientist is so minimal as to be insignificant. Even if such a contention were true (and it is an especially

---

[42]Grene, Knowing and Being, p. 195.

[43]The "inexact non-mathematical knowledge" to which Polanyi refers is the scientist's heuristic vision, a vision that continually occupies a subsidiary position.

[44]Polanyi, The Tacit Dimension, p. 6.

tenuous connotation within the context of discovery), the fact remains that the individual scientist has committed himself to a vision of reality that consistently integrates the given procedures and yields their conclusions tenable. Were he not intellectually satisfied with their ability to depict reality as his own vision would have him understand, the procedures or theories would appear misdirective or ineffectual and, therefore, inappropriate. The final determination of the value of a formal procedure is an act of tacit, as opposed to formal, inference; it requires that scientists informally assess the procedure or theory by estimating its capacity to integrate phenomena in accordance with his heuristic vision.

Let us now analyze Polanyi's conception of "tacit inference." We may distinguish an act of tacit inference from a formal logical operation by indicating that while the latter deals solely with wholly specifiable elements, the former synthesizers unformalizable subsidiary elements. Polanyi explains that the

> ...difference between a deduction and an integration lies in the fact that deduction connects two focal items, the premises and the consequences, while integration makes subsidiaries bear on a focus.[45]

The specific character of the elements of a logical operation permits one to clearly define their relations.

---

[45]Polanyi, American Psychologist, p. 32.

Any operation can be critically reviewed; the processes leading to a conclusion can be reduced to a series of explicit functions and the conclusions can be so analyzed as to systematically reveal their primary assumptions. Polyani writes that, "We can go back to its [the logical operation's] premises and go forward to its conclusions, and we can rehearse the process as often as we like."[46] He contends that such "reversibility"[47] is peculiar to formal logic and not to be found in acts of tacit inference.

An act of tacit inference employs elements that are inexplicit; the relations of these elements bear the same indistinct character. The integration of the subsidiary element in an act of tacit inference may well conclude in the formulation of an explicit idea or principle, but the process itself is not reducible to a series of critically analyzable functions.

Beside the fact that subsidiary elements are, by definition, indistinct as focal objects, they serve to direct a scientist's thinking rather than to act as specific content for analysis. The vectorial quality of subsidiaries, that quality that enables them to function in the integrative process, cannot be understood unless the subsidiaries occupy the subsidiary position in the tacit knowing triad.

---

[46]Polanyi, _American Psychologist_, p. 32.

[47]Polanyi's concept of "reversibility" is largely based on the work of Jean Piaget. See _Psychology of Intelligence_, London, 1950.

One could not, in other words, recreate the vectorial quality of the subsidiary elements a knower integrates by analyzing the conclusions the knower reaches. It may well be possible to deduce various aspects of the premises employed in a tacit inference from an analysis of a given conclusion, but the "logic" underlying their integration would not be reducible.

Examples of such irreducibility and ultimately irreversibility, are readily available in the history of scientific discovery. The observations and mathematical tools employed by Copernicus were available to astronomers for generations before him. It was Copernicus, however, who integrated these observations and mathematical conceptions into a new conceptual whole, a single new focus. Similarly, Newton's discovery of universal gravitation was not preceded by a series of perplexing observations. Newton's genius lay in his ability to synthesize the available observations into a unified conceptual focus.[48] In yet another case, Polanyi notes that Einstein's discovery of relativity theory was "unaided by any observation that had not been available for at least fifty years."[49,50] Furthermore, as stated previously, Popper

---

[48]A similar position is presented by David Bohm in "On Insight and Its Significance for Science, Education, and Values" in Teachers College Record, 80, 3 (1979):406-7.

[49]Polanyi, Personal Knowledge, p. 11.

[50]Einstein confirmed Polanyi's argument when, responding to an inquiry Polanyi had made regarding the discovery

concludes that Einstein's discovery of relativity cannot in any way be directly attributed to a systematic analysis of the observations of that time. Einstein integrated the empirical information in accordance with a new vision of reality that formed his conceptual focus.

The point of these examples lies in the fact that they demonstrate that although discoveries, once made, may be seen as a logical or at least systematic extension of preceding theories, the process of discovery cannot be broken down into a step-by-step progression leading from previous theories to ones newly discovered. The forward movement, the leading edge of scientific inquiry, is not guided by formal, specifiable processes. The integrations of entities, the tacit inferences, that guide the scientist in his expansion of his knowledge cannot be traced back to their premises in such a way that one could state in an authoritative manner why the scientist chose to proceed as he did. We may conclude that scientific knowledge expands by way of tacit inferences that cannot be reduced to analytic operations nor traced back to their premises.

One might respond that such arguments reinforce Popper's notion that scientific discovery and assessment can be separated. It appears, one might contend that Polanyi classifies intellectual operations into two categories: one reducible

---

of relativity theory, he stated that the Michelson-Morley experiment—believed by positivists to have provided the basis for Einstein's work—had a negligible effect on his research. Polanyi, Personal Knowledge, p. 101.

and critically available and one irreducible and a-critical. The former classification seems equivalent to the operations that could be employed to assess knowledge while the latter classification seems to include those operations utilized in acts of discovery. Why then should Polanyi not separate his theory of discovery from his theory of assessment?

Polanyi's response is multifaceted and has already been developed with reference to a variety of issues. His concept of the fiduciary character of science, commitment, and the metaphysical referent of scientific theory preclude the possibility that science could be framed in purely logical terms. With regard to this particular issue, however, Polanyi maintains that tacit inferences underly the application and evaluation of given scientific theories. The question of where to seek refutations, how they can be extracted, and when the evidence constitutes a valid refutation require personal and informal judgments. The tacit application of one's presuppositional network, one's heuristic vision, is as necessary to the assessments of scientific theories as it is to the assessment of the reality underlying the perceived configurations of data. Polanyi writes,

Let us recognize that tacit knowing is the fundamental power of the mind, which creates explicit knowing, lends meaning to it and controls its uses. Formalization of tacit knowing immensely expands the powers of the mind by creating a machinery of precise thought, but it also opens up new paths to intuition; any attempt to gain complete control of thought by explicit rules is self-contradictory... The pursuit of formalization will find its true place in a tacit framework.

> In this light, there is no justification
> for separate approaches to scientific ex-
> planation, scientific discovery, learning
> and meaning. They ultimately rest on the
> same tacit process of understanding.[51]

Let us return to our inquiry into the nature of
tacit inference. We have thus far stated that it is an informal process of integrating <u>subsidiary</u> elements into a focal
unity, and that this process of integration can neither be
broken into separate operations nor completely explained by
tracing its explicit representation from its conclusions
back to its premises.

Another characteristic of tacit inference that
distinguishes it from formal logic lies in the fact that it
"is not damaged by adverse evidence, as explicit inference
is."[52]   That is, the integration of subsidiaries can continue to bear a consistent focus even though a given subsidiary may logically contradict such efforts. This occurrence
is possible only because the subsidiaries acting in any given
situation are integrated informally or summarily. Polanyi
explains:

> An integration established in this summary
> manner will often override single items of
> contrary evidence. It can only be damaged
> by new contradictory facts if these items
> are absorbed in an alternative integration
> which disrupts the one previously established.[53]

---

[51]Grene, <u>Knowing and Being</u>, p. 156.

[52]Polanyi, <u>American Psychologist</u>, p. 32.

[53]Ibid., p. 33.

The specific function or significance of any particular subsidiary is relative to its ability to function in concert with other subsidiaries. Where that function is disruptive or simply unfulfilled, the integration can continue with the possibility that the subsidiary may be capable of bearing on the focus in a manner as yet unanticipated.

Polanyi illustrates this point by referring to a psychological experiment in which subjects were shown a skewed room in which a small boy standing in one corner appears to be taller than a grown man standing in a second corner. The incongruity of the relative heights of the man and the boy did not prevent the subjects from seeing the skewed room as rectangular. Only after a variety of clues were available to the participants (such as being allowed to tap the walls with a stick) did they perceive the room as non-rectangular. The optical illusion could only be dispelled by the presentation of a pattern of contradictory clues that could support a new integrational effort.

This characteristic of acts of tacit integration is reflected in the significance of anomalies in the progress of science. Polanyi cites a number of cases where anomalies have not succeeded in refuting theories and where "the stability of theories against experience is maintained by epicyclical reserves which suppress alternative

conceptions in the germ...."[54,55] An individual
anomaly does not necessarily refute a scientific
theory as the theory may some day be found capable of
consistently integrating the observation on the basis
of further evidence or a refinement of the theory itself. Only if a number of anomalies can be integrated
into a new conceptual model or if they become so problematic that the integrational structure in use dissolves
can the dismissal of a theory be affected.

The final characteristic of tacit inference that
distinguishes it from formal logical processes lies in
the fact that the subsidiary pole of the tacit triad is
a complex, largely undifferentiated term. Its bearing
upon a focus does not proceed in a prescribed fashion
where items are classed and operations performed in

---

[54] Polanyi, _Personal Knowledge_, p. 292.

[55] As noted previously, D.C. Miller found evidence
inconsistent with Einstein's relativity theory only to
have the evidence set aside on the grounds that the evidence would one day turn out to be consistent with the
theory or simply invalid (_Personal Knowledge_, pp. 13,
14). In a second case, variances in planetary motions
inconsistent with the prediction of Newtonian gravitation were observed for 60 years before the discovery of
Neptune and were given little heed by astronomers as they
hoped the evidence would eventually be found consistent
with Newton's expectations (_Personal Knowledge_, p. 20).
In still another case, despite numerous contradictions
to Arrhenius' theory of electrolyte dissociation, "Scientists were satisfied with speaking of the 'anomalies of
strong electrolytes' without doubting for a moment that
their behavior was in fact governed by the law that they
failed to ovey." (_Personal Knowledge_, p. 292) Not until a
viable alternative conception was offered were significant

series. Acts of tacit inference synthesize numerous elements from a variety of sources in a dynamic process where all the subsidiaries are mutually transformed and interwoven. This aspect of the integrational process can be a source of misunderstanding and confusion.

Let us consider the simple act of perceiving the man and the boy in Ames' experiment; we may say that the participants were subsidiarily aware of (1) the minute adjustments made by their eyes to bring the scene into focus, (2) the background of the skewed room, and (3) their past experiences with the relative sizes of men and boys as well as the rectangular nature of most rooms. The question arises as to how these items could be synthesized to yield a clear focus on the relative heights of the man and the boy.

The notion of a series of operations runs into logical and practical difficulties. Firstly, if we say that one is subsidiarily aware of one's own eyes to bring the room into focus and then one employs the background of the room as a subsidiary to focus on the boy and the man, etc., then we make individual objects both subsidiary and focal elements in their bearing upon a

revisions of Arrhenius' theory made (<u>Personal Knowledge</u>, pp. 292-293).

single focus. The subsidiary/focal distinction would thus be diffused.

Secondly, Polanyi has consistently maintained that past experience plays into our everyday perceptions. He states,

> In passing from the visionary contemplation of an object to its observation, we do make an affirmation....This is an act involving a commitment....It establishes a conception of reality experienced in terms of a subsidiary awareness of the coloured patches which had previously been experienced as such in an act of contemplation.[56]

Once we step beyond the concept of perception as a passive reception of various frequencies of light in patchwork patterns, the act of seeing, the act of observing a dimension of interrelated objects and events, then perception includes the influences of past experiences. In the case of the Ames experiment, for example, it was found that "primitive people who have seen fewer normal rooms are less susceptible to the [illusion perceived by their more sophisticated counterparts]."[57]

In sum, tacit inference appears to be a dynamic process in which subsidiaries from various sources are integrated summarily in such a manner that they dynamically interweave and undergo mutual transformation (as their relative functions and relations are altered) to bear upon

---

[56]Polanyi, _Personal Knowledge_, p. 99.

[57]Grene, _Knowing and Being_, p. 165.

a single focus.

This chapter has explored Polanyi's concept of the subsidiary/focal nature of the process of knowing. However, we have not yet analyzed the all important third element in the tacit knowing triad: the active person. Further, the question must be asked, how a presuppositional vision can, in part, become explicit and formal. Chapter III will explore both these questions. In addition, the question of how subsidiaries can function as an ineffable framework in the creation and utilization of explicit scientific theories still remains. Chapter IV will specify Polanyi's explanation of the mechanisms scientists employ in the discovery of new and unique theories.

Chapter III

THE KNOWER IN THE KNOWN

We have thus far analyzed Polanyi's notion of the metaphysical referents of science and his concept of the unformalizable, tacit component of scientific knowledge. We also clarified the necessary relation between Polanyi's metaphysic and the indeterminant nature of scientific knowing. The general structure of Polanyi's concept of the process of knowing had three components: a focal object (C), a set of subsidiaries (B) used to attend to the focal object and a knower (A) who brings the subsidiaries to bear upon the focus.

We have not, however, clarified Polanyi's concept of the knower. Is the knower separate from the subsidiaries he uses? If the knower is distinct from and operates in accordance with principles unattributable to the functioning of subsidiaries, where do the principles that guide the knower come from? What is their metaphysical status? In essence, we have yet to distinguish the ontological status of the person in Polanyi's "personal" knowledge.

It is the purpose of this chapter to critically analyze Polanyi's concept of the mind, its relation to the body and its role in the process of discovering and accessing articulate scientific theory.

Polanyi's concept of "indwelling" provides us with a bridge between the general structure of the tacit knowing triad and the nature of the individual in that triad. We shall systematically follow the implications of "indwelling" from the mind/body relation in simple physical acts to the ontological status of the mind itself in scientific conceptualization. This does not, of course, imply that the mind/body relation or the ontology of mind undergo fundamental transformation in such tasks but only that different perspectives can be gained through the graded analysis of physical and temporal activities. Let us begin with a brief review of the tacit knowing triad.

The tacit knowing triad, as described in Chapter II, consists of a knower (A) who focuses on an object (C) through his subsidiary awareness of yet other objects (B). The nature of the subsidiaries (B) varies dramatically with the activity being performed. A craftsman may employ his tools to attend to the creation of a work of art; an infant may manipulate the muscles of his eyes to focus upon a visual object, or a scientist may employ mathematical processes in the discovery and assessment of a scientific theory. This is not to say that such persons are limited to these specific sets of subsidiaries; they are not. The scientist, for example, not only employs a particular statistical method in his research but also subsidiarily applies particular theories[1] and his general heuristic vision as a background upon which the object being studied appears as a figure.[2] As a scientist's commitments to specific scientific theories and statistical procedures change, his subsidiary intellectual background is altered and the object studied assumes new intellectual significance, new scientific status. We must be careful to note, however, that a scientist's subsidiary intellectual background is not equivalent to his mind. The mind <u>uses</u> subsidiaries; we are given no reason to believe that the mind is composed <u>of</u> subsidiaries.

The critical factor for our present purposes lies in the fact that the subsidiary/focal character of the knowing process tends to place the knower within the context of the subsidiaries (B)--whether they be the muscular actions in an infant's eyes or the highly complex intellectual structures employed in the process of inquiry--looking <u>out</u> at the object being considered (C). Subsidiaries form an unformalized existential dwelling, as it were, from which the knower attends to an object. The existential experience of the subsidiary elements of knowledge need not be abandoned when we consider scientific knowledge. We have already seen that Polanyi clearly rejects hard and

---

[1]Theories are here said to be subsidiarily applied as they are selected and applied with a personal grasp of their implications and functions.

[2]Historically, one can see a similar perspective in

fast distinctions between skill and knowledge and between the logical and psychological aspects of cognition. Polanyi's concept of the subsidiary/focal nature of knowledge necessarily actively involves the knower in the perception and conception of the known. (His concept of universal intent prohibits caprice and implants responsibilities transcending the scientist's subjective desires and goals.)

Polanyi further explains,

> ...any particular indwelling is a particular form of mental existence. If an act of knowing affects our choice between alternative frameworks, or modifies the framework in which we dwell, it involves a change in our way of being.[3]

The existential aspect of tacit knowing does not preclude the rigorous use of logical principles and statistical methods. Rather, it is Polanyi's contention that such public and impersonal mechanisms are determined by an individual's existential experience of reality. "There is present [within acts of knowing] a personal component,

_____

the work of Wilhelm Dilthey. Dilthey maintained that all scholarly practice must occur within and take account of particular cultural perspectives and standards. However, Polanyi and Dilthey differ with respect to transcendental motions. Whereas the latter interprets all phenomena within an historical matrix, the former maintains that there are operational principles at work in human beings which transcend culture, e.g. the intellectual desire for coherence.

[3]Marjorie Grene, Knowing and Being (Berkeley, CA: University of California Press, 1974), p. 134.

inarticulate and passionate, which declares our standards
of values, drives us to fulfill them and judges our performance by these self-set standards."[4]

The question arises as to the function of observation
in the subsidiary/focal structure of scientific knowing.
In order to respond to such a question we must keep in
mind that the referents of science are ontological and cannot be wholly defined by that which has been observed.
To this extent, that observations direct a scientist toward
a yet hidden (observationally unavailable) coherent entity,
empirical data act subsidiarily. In Polanyi's terms, it is
by dwelling in them as they puzzle us and maintain our attention that we come to know the coherent ontological
principles governing the observed. Polanyi writes, "All
understanding is based on our dwelling in the particulars
of that which we comprehend. Such indwelling is a participation of ours in the existence of that which we comprehend."[5]

Thus, Polanyi presents himself with a rather complex
dilemma. He has created a situation in which he is forced
to explain how objects independent of our bodies can assume
existential status. Yet more fundamentally, he must offer
a consistent analysis of the question of the mind-body

---

[4]Michael Polanyi, Personal Knowledge: Toward a Post-Critical Philosophy (New York: Harper and Row, 1964), p. 195.

[5]Ibid., p. X.

relation.

Polanyi responds by informing us that the mind is to the body what a coherent principle is to a perceptual field. That is, the former can only be apprehended through a heuristically guided rather than reductionistic integration of the latter. This unformalizable process, Polanyi claims, can best be introduced, once again, through an examination of perceptual processes.

Polanyi explains his approach to the mind-body question in the following way:

> ...the relation between body and mind has the same logical structure as the relation between clues and the image to which the clues are pointing. I believe that the paradoxes of the mind-body relation can be traced to this logical structure and their solution be found in the light of this interpretation.[6]

He further states, "by elucidating the way our bodily processes participate in our perceptions we throw a light on the bodily roots of all thought, including man's highest creative powers."[7]

As we have noted previously, one is subsidiarily aware of the organs of one's body when one uses them to attend to the external environment. As I write upon a page, I rely upon the muscles of my eyes to focus the rays of light entering me through the lenses of my eyes, and I am

---

[6] Grene, Knowing and Being, p. 213.

[7] Michael Polanyi, The Tacit Dimension (Garden City: Doubleday & Company, Inc., 1967), p. 15.

subsidiarily aware of the pressure I feel in my fingertips as I attend to the letters I shape. Thus, I use my body to attend to this physical activity. Polanyi concludes that "our own body is the only thing in the world which we normally never experience as an object, but experience always in terms of the world to which we are attending from our body."[8]

The subsidiary character of our experience of our bodies is dynamically related to our physical interaction with the environment. As we come into contact with our environment, we come to know our bodies as our own rather than an object separate from us. To the extent that we attend to our body as a focal object, we have reduced our existential experience of living therein. Viewing my hand as it crosses this page is little different than a stranger observing it. Of course, I do view my hand as I write, but I do not view it as a focal object. When I view my hand as a focal object it ceases, to that extent, to function subsidiarily, as something I experience as part of my body. Polanyi maintains that even when we focus on an internal pain subsidiary experience usually still predominates.

We may conclude that our bodies are primarily instrumentally known. Polanyi writes:

---

[8]Polanyi, The Tacit Dimension, p. 16.

> Our body is the ultimate instrument
> of all our external knowledge, whether
> intellectual or practical. In all our
> waking moments we are <u>relying</u> on our
> awareness of contacts <u>outside</u> for at-
> tending to these things.[9]

Our bodies have the unique position in the universe of being "the ultimate instrument of all our external knowledge," but their subsidiary and instrumental character can be shared by many physical and intellectual objects. Polanyi maintains that external objects, such as tools or scientific theories, can be subsidiarily known and made to function to shape our intentions practically and theoretically. "In this sense, then," he writes, "to make something function subsidiarily is to interiorize it, or else to pour one's body into it."[10] Polanyi holds that in the use of an object which serves an instrumental function the knower includes it in his subsidiary awareness and existential experience of dwelling in the world.

Let us, once again, consider the act of writing upon this page. While it is true that the path my pen will trace is subsidiarily determined by my eyes and hand,[11] I am also subsidiarily aware of the feel of my pen on the

---

[9]Polanyi, <u>The Tacit Dimension</u>, p. 16.

[10]Michael Polanyi, "Logic and Psychology," <u>American Psychologist</u>, 23, 1 (January 1968):33.

[11]We should also note that numerous other subsidiaries play into the act of writing such as my instrumental awareness of letters, syntax, grammar, language, etc.

page. I am not presently focusing on my pen but attending
to it subsidiarily as the letters form on the page. If my
pen were to fail to produce the ink necessary to write
I might then focus on the pen itself. In so far as it fulfills its function I extend my subsidiary awareness of my
fingertips into by subsidiary awareness of my pen as it
touches the page. It must be noted, however, that I do not
create a string of subsidiary/focal relations from eye to
hand, from hand to pen, from pen to ink, etc.; rather, all
subsidiary factors are woven with one another as they are
integrated instrumentally.

Virtually any one of them could be made focal if it
were to fail to fulfill its function; we would then look at
that object through a different complex of subsidiary factors. Polanyi explains that, "to attend _from_ a thing to
its meaning is to _interiorize_ it and that to look instead
_at_ the thing is to _exteriorize_ or _alienate_ it."[12] An
interiorized object functions subsidiarily with the subsidiary complex of our bodies. It is quite easy to say where
my hand ends and my pen begins but they are intertwined as
a single extension of my intent as I attend to writing.
Polanyi concludes:

> We may say that our own existence, which we
> experience, and the world that we observe

---

[12]Grene, _Knowing and Being_, p. 146.

are interwoven here. Bodily being,
by participating subsidiarily in one's
perceptions and actions, becomes a being in the world, while external observations and projects subsidiarily involving one's own bodily feelings become, up to a point, a self-transformation, an existential choice.[13]

The self-transformation to which Polanyi refers
consists in the inclusion of an object into the subsidiary complex of one's body. Such an inclusive process
is not explicit and formal but is accomplished in the
practice of a skillful activity. In the performance of
a skill, Polyani maintains, the individual subsidiarily
incorporates an object into his attempt to focus upon
the object of his intent. The existential choice is one
whereby the individual assimilates an object within his
sphere of activity rather than holds it at a distance from
himself as a separate object. "The process of integration assimilates them [subsidiary objects] to our body
and to this extent deprives them of their character as
external objects."[14]

* * *

Polanyi identifies four aspects of the process
of tacit knowledge which add clarity and dimension to
his arguments concerning the relation of the body and
mind. In attending from a complex of subsidiaries to a

[13]Polanyi, _American Psychologist_, pp. 33-34.

[14]Grene, _Knowing and Being_, p. 184.

focal object we find what Polanyi calls the "functional structure"of tacit knowing. The existential experience of dwelling in our bodies and interiorized objects provides a personal and active nexus from which we can attend to those objects having a focal status. The significance of the functional structure lies in the claim that consciousness is directed. Polanyi claims that all knowledge arises from the active existential context of one's subsidiaries and is directed toward focal objects. Marjorie Grene, in describing this aspect of Polanyi's theory, states,

> The function of my subsidiary knowledge
> is to direct me to the coherent sight of
> my surroundings. This is the functional
> import of tacit knowing: it guides me from
> proximal, interiorized particulars to the
> integration of a coherent, distal whole.[15]

We may conclude that consciousness is not only purposive in its continuous use of subsidiaries to attend to focuses, but is also bi-polar. Polanyi writes,

> All thought contains components of which
> we are subsidiarily aware in the focal
> content of our thinking, and all thought
> dwells in its subsidiaries, as if they were
> parts of our body. Hence thinking is not
> only necessarily intentional, as Brentano
> has thought; it is also necessarily fraught
> with the roots that it embodies. It has a
> from-to structure.[16]

---

[15]Grene, Knowing and Being, p. IV.

[16]Polanyi, The Tacit Dimension, p. X.

To be conscious implies that one attends from subsidiaries to a focus. The focus cannot be brought into
clarity without the use of subsidiaries, and the subsidiaries when unintegrated in bearing on a focus, are
without significance. This does not imply that all
conscious acts can be explicated, but on the contrary,
that consciousness itself has a from-to structure, a
structure whose subsidiary pole is vectorial and whose
focal pole provides an impetus for personal activity.
Robert Innis, in reviewing Polanyi's concept of consciousness, states, "Consciousness is not a form of inner
perception with its attendant duality of perceiver and
perceived, but an inner experience in the mode of indwelling."[17]

The vectorial quality of the subsidiaries is not
accomplished through a formal logical process. Neither
induction nor deduction fix the relations between subsidiary and focal items; subsidiaries are integrated or
synthesized so as to bring an object into focus.* As
stated previously, Polanyi maintains that the informal
vectorial quality of subsidiaries is a result of the
functional relation between subsidiary and focal elements

---

[17]Robert Innis, "Polanyi's Model of Mental Acts,"
The New Scholasticism (Spring 1973):152-153.

*The concepts of induction and deduction will be
examined in Chapter IV.

rather than of a Jamesian fringe-consciousness. He explains:

> When writing a letter I am fully aware of the pen and paper I am using. The fact that I am focusing my attention on these particulars, but attending from them to that which they mean, reduces them to a subsidiary status, but does not render my knowledge of them subconscious or preconscious, or such as one has of an identifiable Jamesian fringe.[18]

The indwelling experience is a function of the subsidiary pole of consciousness rather than of a specific content. To dwell in something simply refers to its subsidiary character in a given activity. This becomes of great significance when we consider the subsidiary quality of our bodies. Polanyi notes, "All acts of consciousness are then not only conscious of something, but also conscious from certain things which include our body."[19] As we dwell within our bodies, as we use them in a subsidiary capacity, our consciousness merges with our bodily existence. Thus it would appear that Polanyi has found a philosophical framework for overcoming Cartesian dualism. We have seen that the mind and the body merge as one dwells in one's body in attending to objects in the environment. In the case of the mind/body question, we are led to conclude that the mind and body together

---

[18]Grene, Knowing and Being, p. 194.

[19]Ibid., p. 214.

give rise to consciousness as we attend to a focal object. Neither the body nor the mind could be said to be a center of consciousness without the other; their integration permits the emergence of this aspect of being. It would appear that the mind and body together form the subsidiary pole of consciousness while external (non-interiorized) objects form the focal pole. We will see later, however, that Polanyi retains his own dualism between body and mind.

The second aspect of the process of tacit knowing identified by Polanyi refers to the notion that subsidiary particulars produce a perceptual ground that gives form to the focal object. This is known as the "phenomenal" aspect of tacit knowing. Subsidiary items in a perceptual field give rise to the particular phenomenal pattern that will be perceived; in science new observations related to a given theory become assimilated into a larger picture of reality.

Polanyi states that through the interaction of subsidiaries, "a perceived object acquires constant size, color and shape; observations incorporated in a theory are reduced to mere instances of it; the parts of a whole merge their isolated appearance into the appearance of the whole."[20] The appearance of comprehensive entities

---

[20]Grene, Knowing and Being, p. 141.

is dependent upon the subsidiary integration of their elements and contexts. Were nonempirical subsidiaries (presuppositions, theories in use, etc.) absent from perceptual acts objects would appear as, at best, vague and diffuse patches of color; were they absent from the observation in scientific inquiry, empirical objects and events would remain equally chaotic and unrelated. As we dwell within our bodies, we are actively involved in the process of perception; as we dwell within the subsidiary context of a scientific theory, we are actively involved in shaping scientific knowledge.

Polanyi distinguishes a third aspect of the tacit knowing process by noting that the meaning of any subsidiary is found in its relation to the whole towards which it points. He refers to this as the "semantic" aspect of tacit knowledge. The words that I write upon this page may be said to carry a significance beyond themselves; they bid the reader dwell on the intellectual framework of this writer as a rational context for the study of a given question in philosophy. Each word, when isolated, loses its significance as it no longer functions to give the reader a clue to this writer's meaning; it is only when the words are considered as a unified complex that they create a meaning--in this case the words carry a presuppositional framework, and the writer asks the reader to assimilate them as a viable context for approaching

the nature of scientific inquiry. Marjorie Grene concludes, "All explicit knowledge, however, crystallized
in the formalisms of words, pictures, formulae, or other
articulate devices, relies on the grasp of meaning through
its articulate forms on the comprehension that is its
tacit root."[21]

The semantic aspect of tacit knowing is not confined
to understanding non-scientific, subjective accounts. As
we noted in Chapter II, understanding a scientific theory
entailed the ability to use it in a vectorial capacity, as
an indicator of coherence in nature. We may now add that
each equation and formal statement has a semantic aspect
in that its meaning transcends the internal structure of
the theory itself. An equation embedded in a scientific
theory implies a preconception of the world which can be
formalized in the equation itself. Thus, equations may
be mathematically correct and yet considered unreliable
and inappropriate for a given field of inquiry, because
they do not provide an intellectual framework in keeping
with the heuristic vision of a scientist.

Further, empirical data have a semantic component.
Machines, for example, embody physical and chemical laws,
but they do so with an overarching purpose that cannot
be described in physical and chemical law. As explained

---

[21]Grene, Knowing and Being, p. XII.

in Chapter II, an army of physicists and chemists could never determine the function of a simple wrist watch. Similarly, Polanyi asserts living organisms also embody physical and chemical laws that are governed by higher organizational or operational principles that defy physio-chemical explanation.[*] Physicists and chemists, for example, could not differentiate healthy tissues from diseased ones or even live cells from dead cells were not the empirical data assimilated in a subsidiary complex of ideas that included the vague notions of health and life itself. On one level life may be described as a property of carbon compounds, but such a conception fails to recognize that life could not be differentiated thusly were not some "instructive" surmise made that such a property existed. Physics and chemistry cannot know of life or death, of microscopic wars, of even the simplest concepts of success or failure. My blood count does not merely state that a given number of white cells are in a certain volume of blood, it indicates possibilities of disease, of possible dangers and risks. Polanyi concludes that,

> ...to know an organism is to acknowledge
> the existence of an individual and appre-
> ciate its correct growth, form and function,
> these features being judged to be healthy
> or abnormal by standards which we consider

---

[*]For a more detailed analysis of this issue refer to the discussion on the irreducibility of comprehensive entities on pages 129 to 137.

apposite to an individual as a member of
its species.[22]

Thus, physio-chemical data are useful only to the
extent that physiologists formulate their research "in
terms of previously known or surmised operational
principles."[23] The semantic aspect of tacit knowledge
indicates that that which is observed often has meaning
beyond itself; the meaning that the observed embodies is
part of an overarching coherence. Without a presuppositional vision that can serve in a heuristic fashion to
parallel, in the intellectual framework of the scientist,
the operational principles in the coherence, physiology
could not progress towards understanding living
organisms.

The fourth aspect that Polanyi distinguishes refers
to that coherence towards which semantic indicators
point. This aspect of the tacit knowing triad refers
specifically to the coherences that govern that which is
seen whereas the semantic aspect of the triad refers to
the meaning of individual objects within a coherent framework. This ontological aspect denotes that the object
being researched is indeed a real object and should,
therefore, be expected to reveal itself in new, and as yet,

---

[22]Michael Polanyi, The Study of Man (Chicago:
The University of Chicaogo Press, 1958), p. 74.

[23]Ibid., p. 54.

unexpected ways. The important thing here is to remember that the object itself is not necessarily a tangible entity but can be an operational principle. As such, when empirical conditions are altered or viewed from a new perspective consistent, yet not necessarily analytically predictable results should ensue. The ability to predict the behavior of the operational principle under new conditions depends not on a vector-like analysis of its past activity, but rather upon the scientist's ability to dwell within that operational principle as his own intellectual ground. The difference between a vector analysis and an indwelling experience of given data lies in the fact that the former operates within the scope of preciously accepted relations while the latter incorporates informal premises and skills capable of anticipating wholly new, analytically unpredictable phenomena.

Certainly, it may be argued that part of the fruitfulness (however it may be defined) of a scientific theory refers to its ability to bear consistent implications in new inquiries. Polanyi agrees but asserts that the elaboration of the implications of a given theory is not usually a wholly deductive process. As noted previously, Polanyi maintains that scientific theories are founded on fiduciary presuppositions that constitute a heuristic vision. Such vision, such perspective, is embodied in each and every scientific theory. Scientists utilizing a given theory dwell, as is were, in the perspective it offers; they approach new areas of inquiry from its (the theory's) intellectual vantage point. Where scientists find that their personal vision is in conflict with the heuristics of the scientific theory they are studying, they may attempt to refute, refine or reject them. Grounded in the perspective offered by a scientific theory that is in primary agreement with a scientist's own sense of truth and reality, a scientist may apply such insights to new conditions or whole fields of research to reveal new and unique aspects of the object he studies. Copernican theory provided the intellectual dwelling place for Kepler to study the motion of Mars. Kepler's law of planetary motion provided the foundation for, but did not directly imply, Newtonian gravitational principles. Newton provided the ripe question for, but did systematically indicate, Einstein's relativity theory. The non-deductive implications of the questions borne of the Einsteinian vision of reality are first being discovered. The Nobel prize winning physicist, David Bohm, explains in his latest book, <u>Wholeness and the Implicate Order</u>, that the most sophisticated inquiries in physics reveal that even subatomic particles embody an implied order that can only be grasped through a recognition of an inexplicit comprehensive unity. Thus, observations in physics are said to have an overarching coherent aspect that extends beyond the observations themselves. Bohm

writes,

> The idea of a separately and independently
> existent particle is seen to be, at best, an
> abstraction furnishing a valid approximation
> only in a certain limited domain. Ultimate-
> ly, the entire universe (with all its 'partic-
> les,' including those constituting human
> beings, their laboratories, observing instru-
> ments, etc.) has to be understood as a
> single undivided whole, in which analysis
> into separately and independently existent
> parts have no fundamental status.[24]

Polanyi similarly describes the ontological aspect

of tacit knowledge,

> The implications of new knowledge can never
> be known at its birth. For it speaks of
> something real, and to attribute reality to
> something is to express the belief that its
> presence will yet show up in an indefinite
> number of unpredictable ways.[25]

Heuristic frameworks thus provided by antecedent

theories yield subsidiary contexts for contemporary

scientists to attend to pertinent questions about and clues

to yet hidden aspects of reality. This is not to say that

the foramlisms such as the mathematics of Copernican

theory or Einstein's relativity theory lack specific im-

plications but rather that as such systems are employed

their internal workings must be applied in such a way as

to parallel the generative operational principles actually

---

[24] David Bohm, _Wholeness and the Implicate Order_ (London: Routledge and Kegan Paul, 1980), p. 174.

[25] Polanyi, _Personal Knowledge_, p. 311.

functioning at the ontological level. They should, if they operate in such a fashion, create a new problem and questions leading to their own further specification or transformation through new contacts with reality.

In any case, the explicit element of scientific theories is dependent for its scientific status on the presuppositional commitments of the scientific community and individual scientists. To the extent that scientists accept a theory as a viable intellectual dwelling place, the presuppositional commitment is made (even though it might well turn our to be misguided) that the theory reflects in its own structure and form the operational principles governing natural events. As a general rule, we should expect that as the presuppositions and dependent explicit theoretical statements of science mirror operational principles, they are imbued with heuristic power. Polanyi concludes,

> Scientific discovery reveals new knowledge, but the new vision which accompanies it is not knowledge. It is _less_ than knowledge, for it is a guess; but it _is more_ than knowledge, for it is a foreknowledge of things yet unknown and at present perhaps unconceivable....In fact, without a scale of interest and plausability based on a vision of reality, nothing can be discovered that is of value to science....[26]

The presuppositions underlying scientific inquiry play a subsidiary role. Scientists attend _from_ their

---

[26]Polanyi, _Personal Knowledge_, p. 135.

dwelling places within their visions to the data or theory being studied. When a theory is being used it also assumes a subsidiary position. We must reiterate, however, that dwelling in both a theory and its presuppositions does not imply a series of subsidiary/focal relations. The subsidiary pole of consciousness is not explicit or logical; it is a complex matrix, functional and informal by nature.

To understand a given area of research, a scientist must dwell within both the subsidiary context provided by the heuristic vision he has assumed and, in cases where a particular theory is being applied, the vectorial quality of the scientific theory. Such indwelling will be fruitful to the extent that the subsidiaries function in the scientist in a manner parallel to the operational principles functional in the ontological field. Expectations derived from the former should parallel the continuity of the latter in areas yet to be researched.

*   *   *

The question now must be asked as to how we can begin to understand the ontological referent of another human being. The sheer analysis of a human being's behavior in relation to environmental factors alone would not reveal his operational principles (whether we are focusing on individual or species-wide characteristics). Just as the teleological aspects of a

machine or an organism's tissues would not be evident to
physicists and chemists, Polanyi maintains that the intentional or purposive elements of human behavior are precluded from behavioral analysis. Polanyi explains,

> Dwelling in our body, clearly enables us to
> attend <u>from</u> it to things outside, while an
> external observer will tend to look <u>at</u>
> things happening in (or to) the body....He
> will miss the meaning these events have
> for the person dwelling in the body and
> fail to share the experience the person has
> of his body.[27]

Polanyi claims that behavioral analysis is incapable
of integrating a person's behaviors from without as they
are subsidiarily integrated by the person himself. Each
behavior has a semantic aspect, a significance beyond itself that results from its functional relation to the
whole scope of the person's existential experience. The
situation mirrors the attempt to understand music by
analyzing the frequencies, amplitudes, and intervals of
the sounds being produced; the principle, integration,
and weaving the musical tones would be lost. Similarly,
the behavior a person exhibits is dependent upon purposes
and intentions unified in a complex subsidiary network
within the individual. Polanyi maintains that it is not
the environment that organizes behavior but rather the
person who dwells within the body, experiences the input

---

[27]Grene, <u>Knowing and Being</u>, p. 148.

of the environment, and purposively directs his own responses. In order to understand such determinants one must recognize their ontological character.

The perception of objects requires a subsidiary complex to make the perceptions meaningful: to unite them in time and space with other objects and events. The response to the stimuli encountered in the environment is based upon such interpretation and cannot be understood except in terms of this context. As the environment becomes more complex and as one's capacities expand, the reductionistic analysis of the relation between environment and behavior becomes increasingly ineffectual. The apparent correspondence between behaviorist theories and empirical evidence is less the result of scientific rigor than of the unnoticed insertion of teleological presuppositions in behaviorist terminology. Polanyi explains,

> ...even the most elaborate objectivist nomenclature cannot conceal the teleological character of learning and the normative intention of its study. Its supposedly objective terms still do not refer to purposeless facts but to well functioning things. Something is a "stimulus" only if it succeeds in stimulating. And though "responses" may be meaningless in themselves, the state of affairs called "reinforcement" functions as such by converting at least one particular response into a sign or a means to an end.[28]

---

[28]Polanyi, Personal Knowledge, pp. 371-372.

Failure to acknowledge and understand the semantic and ultimately ontological significance of each stimulus and response reduces the heuristic value of a given theory relating either to an individual or a species. Since stimuli are interpreted and responses are generated within and from subsidiary contexts, a cohesive interpretation of their interaction would require that the researcher's presuppositions and theories provide parallel interpretative and generative structures to predict future behaviors. The predictive value of behaviorist theories results largely from the tacit assumptions and conceptual integrations researchers make to fill the gap between what is observed and the structure that is used to interpret the same. Polanyi concludes,

> The present psychology of learning which strives for objectivity, achieves a semblance of its aim as follows. First it curtails its subject matter to the crudest forms of learning. Second, it exploits the ambiguity of its supposedly impersonal terms, so that they will apply to the actual performance which is covertly kept in mind.[29]

In the last analysis, Polanyi asserts that behaviorally circumscribed psychology, with its emphasis on the empirical parameters of scientific theory, precludes the study or very conception of a human purpose or a human mind that can effectively intervene between a stimulus

[29]Michael Polanyi, "The Body and Mind," The New Scholasticism XIII, 2 (Spring 1969):204.

and a response. The behavioral correlations established refer only to phenomenal patterns, that is, coincidental conjunctions of stimuli and behavior, and therefore lack the implicit predictive power of a theory mirroring onto-logical principles.

Polanyi's acceptance of the semantic and ontologi-cal aspects of human beings places him in a position where he attributes purposiveness to human activity and main-tains that such purposiveness is conferred by a human mind. We thus arrive at the question of the mind/body relation from a new angle. It had appeared before that Polanyi has, perhaps, provided a framework to overcome a mind/body dualism in his concept of indwelling. The ques-tion must now be more directly addressed to determine the ontological status of the mind itself. Polanyi recognizes this implication and responsibility when he writes, "Our theory of knowledge [tacit knowledge] is now seen to im-ply an ontology of the mind. Objectivism requires a spe-cifiably functioning mindless knower."[30]

One may object to Polanyi's assessment of behavior-ist theory and his search for an ontology of the mind by asserting that intelligent behaviors are the actual work-ing of the mind rather than results of its activity. Gilbert Ryle, for example, states that the mind and

---

[30]Polanyi, Personal Knowledge, p. 264.

body are 'not two things,' 'not tandem operations';
[they] contain' no 'occult causes,' 'no occult antece-
dents,' no 'ghost in the machine.'[31] This argument can
be reduced to the fundamental concept that behavior is
the result of physiological and neurological processes
rather than "irreducible mental intentions." In short,
it may be said that the mind is composed of electro-chemical processes, electro-chemical processes that constitute the mechanisms whereby behaviors are affected.
Thus, in Ryle's words, "'most intelligent performances
are not clues to the mind; they are those workings.'"[32]

Polanyi, in responding, clarifies his concept of
indwelling by specifying the nature of the subsidiary
factors in the existential experience of our bodies. We
had previously noted that in the act of perceiving an object we rely upon numerous subsidiaries such as the
muscles in our eyes. The question now before us is
whether or not we are subsidiarily aware of the neurons
that activate the muscles of our eyes, and similarly, if
we are subsidiarily aware of the neurons that bear impulses produced by light striking the retina. Questions
such as these would normally be far beyond the scope of
philosophic discourse, but our present task is less to

[31]Grene, Knowing and Being, pp. 222-223.

[32]Polanyi, American Psychologist, p. 34.

explore the physio-chemical structures of the nervous system than to indicate the possible limitations of such inquiries. Polanyi maintains that empirically circumscribed scientific research precludes a full understanding of the human mind and its workings as he maintains that it (the human mind) exists as a coherent entity within which electrochemical mechanisms are subsidiary factors, factors which act as mere clues to the mind itself.

Polanyi states that he believes that the subsidiary character of the body we have thus far developed extends into the tissue of the brain itself. He states, "I suggest now that we extend this scheme to include a subsidiary awareness of cortical processes, by recognizing that we are aware of these too in terms of that which we perceive."[33] The subsidiary character of cortical processes derives from their functional relation to the perceived object. The electro-chemical patterns that occur within cortical tissues carry impulses that have a meaning and significance beyond their chemistry. Just as my speech carries messages that elude the inquiries of researchers focusing on the sound that comes from my mouth, the tissues of the cortex perform functions (in the act of focusing on an object) within a context that

---

[33]Polanyi, The New Scholasticism, p. 199.

is unavailable to physical and chemical concepts. Such a context is illustrated by the universal intent under-lying scientific inquiry itself.

The connection between various elements in a line of reasoning, the balancing of one's judgment, the strength of one's intellectual commitments, the selection of scientific pursuits, etc. are, according to Polanyi, functions of intellectual intentions and not purely elec-tro-chemical occurrences. Polanyi, himself a medical doctor, notes that "Cortical traces spread simultaneously along many lines that have no anatomical link to any single point in the brain, and thus the unity of consciousness has no brain structure."[34]

Thus, Polanyi's concept of mind transcends physio-logical processes. The data supplied by the senses and carried over to the neurons must be interpreted by the perceiver. Raw sensory data is teleologically processed; the perceiver does not merely receive light impressions that focus themselves or stimulate cortical tissues in the speech center causing subjects to verbally identify ob-jects. The perceiver must focus his eyes, the perceiver must integrate the raw sensory data within a matrix of expectations and make judgments about what indeed has been seen. In short, the subsidiary/focal structure of

---

[34]Polanyi, _American Psychologist_, p. 39.

consciousness necessarily produces teleological considerations whereby subsidiary factors are integrated to produce a meaningful and coherent focus. The quest for
rational coherence, whether it be perceptual or conceptual, is teleological.

The view that the nervous system functions solely
on the basis of physio-chemical laws precludes the quest
for rational coherence on the part of a subject. Polanyi
maintains that by focusing on cortical processes neurolo-
gists loose sight of the fact that such mechanisms are
subsidiary factors experienced by the subject and used
to fulfill a quest for rational coherence.

Focusing on subsidiaries deprives them of their
vectorial character and renders the operational principle
working through them unavailable. For example, when
focusing upon the electrical impulses that flow over a
telephone wire one could never discover that the electrical charges carried meanings and varied with purposes
that were not within the wires themselves. Studies could
be made correlaing electrical frequencies and time intervals and perhaps descriptive theories could be produced.
Without the notion that the telephone wires are part of
a machine whose purpose it is to transmit meaningful
symbols, physicists could never understand the principles
governing its operation or behavior. Similarly, the
physio—chemical study of cortical processes could never

reveal their function or significance, their meaning to the experiencing subject. Polanyi states, "...a neuro-physiologist, observing the events that take place in the eyes and brain of a seeing man, would invariably fail to see in these neural events what the man himself sees by them."[35]

Polanyi maintains that the overarching determinants of thought extend beyond cortical functions to the inte-grative efforts of the individual who existentially dwells within them. Contemplative, as opposed to electro-chemi-cal, factors that shape thought include, "the capacity for understanding a meaning, for believing a factual statement, for interpreting a mechanism in relation to its purpose, and on a higher level, for reflecting on prob-lems and exercising originality in solving them."[36] The contemplative context, the quest for rational coherence and the whole realm of personal judgment provide the existential framework for the integration of sensory data and cortical processes. Polanyi explains,

> ...neural functions supply...signs but they do not supply the interpretation. Since this interpretation forms no part of the nervous system, the system cannot be said to feel, learn, reason, etcetera. These are experiences or actions of the subject using his own neural processes.[37]

---

[35]Grene, Knowing and Being, p. 224.

[36]Polanyi, Personal Knowledge, p. 263.

[37]Polanyi, The New Scholasticims, p. 40.

Polanyi concludes that the operational principles of the mind are of a different order than physiological processes; they are not two aspects of the same thing. "If mind and body were two aspects of the same thing, the mind could not conceivably do anything but what the bodily mechanism prescribed."[38] Polanyi states that his theory of knowledge retains the dualism of mind and body.[39] Previously Polanyi's theory of indwelling seemed to unify both the body and mind in a subsidiary/focal context. The body seemed to be imbued with conscious intention and all thought seemed to be bodily-rooted. It now appears, however, that Polanyi holds that the mind and body are separable entities. The body performs a subsidiary function and to that extent may be said to be a dwelling place, but it is no more ontologically united with the mind than a tool is with the hands of a crafts-man. How then does Polanyi conceive of the mind/body re-lation when he appears to supply frameworks for both refutation and reaffirmation of a mind/body dualism?

Polanyi supplies us with a clue to his position when he states, "We may say..., quite generally, that where-ever some process in our body gives rise to conscious-ness in us, tacit knowledge will make sense of the event

---

[38] Polanyi, American Psychologist, p. 40.

[39] Ibid., p. 34.

in terms of an experience to which we are attending."[40]
Polanyi thus subscribes to a rather unique dualism in
which the body may be said to supply the impetus of consciousness and the mind may be said to supply its form.

The tacit knowing <u>triad</u> is not a tacit knowing <u>diad</u>
even though Polanyi's fundamental premise is that the
process of knowing involves bringing subsidiaries to bear
on a focus. The third element in the triad is the knower
himself, an element separate from the complex of entities
occupying a subsidiary position. The body, in occupying
part of the subsidiary pole of consciousness (that is,
in providing a point of "insertion" into the world), provides an existential bridge through which the phenomena
of the world, focal entities can flow into man's mind.
The mind itself, however, occupies the all important
third position in the triad. It stands above subsidiaries and focuses. Both the mind and body interact in
acts of consciousness for, as stated previously, "Acts
of consciousness are...not only conscious of something,
but also <u>from</u> certain things which include our body."[41]
This is not to say, however, that the mind is subsumed
in the subsidiary/focal relations, the from/to structure
of consciousness. It is rather to say that when the

---

[40]Grene, <u>Knowing and Being</u>. p. 147.

[41]Ibid., p. 214.

mind is actively engaged in acts of consciousness, <u>it</u> <u>uses</u> subsidiaries to attend to a focus.[42] The mind, in overseeing the tacit integration of subsidiaries, including cortical processes, operates according to principles transcending those of neurophysiology (the body). Two dinstinct levels of operation take place in acts of consciousness. The first level refers to subsidiary physiological functioning and the second refers to the overarching contemplative intentions of the mind.

The relation between the two separate levels of operation can be explained in terms of Polanyi's principle of dual control. Let us consider for a moment an object whose mechanism involves a purpose, i.e. the radio beside my desk. On one level it may be said to operate according to the laws of physics and on another level it may be said to operate in accordance with the laws of electrical engineering. It is true that the latter level is dependent upon the former, but the former level does not wholly determine the principles of the latter. The teleological foundations of electrical engineering permit the use of physical laws that could not be used

---

[42]Correspondence between Polanyi's argument and the Kantian notion that the categories and schemata of the mind order sensory impressions are rather limited. Polanyi emphasizes the tacit and irreducible elements of thinking; he provides no categorical structures. Most importantly, Polanyi provides no argument for synthetic a priori processes.

themselves to recognize such aims. Polanyi notes,
"Viewed in themselves, the parts of a machine are meaningless; the machine is comprehended by attending from
its parts to their joint function which operates the
machine."[43] Without the concept of the machine's purpose
the parts of the machine become isolated entities having
only coincidental correspondence with one another. The
laws of physics furnish a wide basis for the creation of
objects that embody purposes that physics cannot comprehend. Polanyi refers to the teleological conditions
beyond the scope of physics "boundary conditions." The
boundary conditions, in this case, are controlled by aims
that employ but are not established by the laws of
physics. Thus, the radio beside my desk embodies two
sets of principles and is subject to the requirements of
each; it is subject to "dual controls."

Polanyi explains,

> Such is the stratified structure of compre-
> hensive entities. They embody a combination
> of two principles, a higher and a lower.
> Smash up a machine, utter words at random, or
> make a chess move without purpose and the
> corresponding higher principle--that which
> constitutes the machine, that which makes words
> into sentences, that which makes the moves of
> chess into a game--will all vanish and the
> comprehensive entity which they controlled
> will cease to exist.[44]

---

[43]Grene, Knowing and Being, p. 153.

[44]Ibid., p. 217.

When Polanyi applies the concepts of boundary
conditions and dual control to the human being, he ar-
rives at the conclusion that the process of thinking includes physiological functions but in an ancillary
fashion (much in the same way my radio employs laws of
physics). Polanyi states,

> Mental principles and the principles of
> physiology form a pair of jointly operat-
> ing principles. The mind relies for its
> workings on the continued operations of
> physiological principles, but it controls
> the boundary conditions left undetermined
> by physiology.[45]

All thought is therefore incarnate because the
neurological principles of the mind provide the mechanisms for the process of thinking. The incarnate aspect
of thought does not, however, provide the contemplative
principles that integrate thoughts with reference to
their content. The notion that thought is incarnate
does not necessarily dispel a mind/body dualism; it
merely indicates that physiological as well as contemplative (independent requirements of the contemplative
mind) processes are involved in the act of thinking.[46]

---

[45]Polanyi, American Psychologist, p. 40.

[46]Polanyi complicates matters by insisting that
explicit inferences are the result of "fixed neural
structures: while tacit relationships depend upon the
"integrative process of the mind" (Knowing and Being, p.
219). This proves troublesome for Polanyi as both reasoning processes incorporate teleological aspects unattributable to the laws of physics and chemistry. We may perhaps
restate Polanyi's conclusion thusly: explicit acts of

Polanyi concludes, "The dualism of mind and matter [is] ...but one instance of the dualism prevailing between every pair of successive ontological levels."[47]

Polanyi's central argument has been that the purposive elements of the mind cannot be reduced to pure physio-chemical mechanisms because physics and chemistry do not provide frameworks for recognizing the existence of purposive structures or for analyzing their effectiveness. Polanyi writes, "The laws of physics and chemistry do not ascribe consciousness to any process controlled by them; the presence of consciousness proves, therefore, that other principles than those of inanimate matter participate in the conscious operations of living things."[48] It is important to note that for Polyani consciousness implies that one attends to an entity through the use of other entities. Thus, Polanyi is presented with the problem of explaining what factors form the basis of teleological aims; that is, he must clarify the ontological foundations of the mind that shape the purposes embodied in acts of consciousness. More specifically, he must respond to the question of what it is in a man's mind that establishes the contemplative requirements of thinking itself. If consciousness is

_____

reasoning may correspond to particular processes, but the cortical processes cannot create the demands that undergird rational thought.

[47]Grene, _Knowing and Being_, p. 155.

[48]Ibid., p. 218.

dependent upon higher operational principles and general aims, it is quite reasonable to inquire as to how those principles were created and those aims set.

Polanyi responds to such inquiry with his notion of emergence. Section IV of Polanyi's Personal Knowledge and several essays (e.g., "Life's Irreducible Structure") present Polanyi's concept of emergence. The scope and complexity of this aspect of Polanyi's work prohibit a comprehensive analysis but a brief account will suit our present needs.

Stratified entities embody two sets of laws that relate in a hierarchical fashion. The lower laws consist of those principles which permit the emergence of the higher level principles. Higher level operational principles consist of select instances of lower level principles. The components of this select group can occur through human intervention--such as in the cases of architecture or electronics--or through random events-- such as, Polanyi believes, in the case of the emergence of living organisms. The higher operational principles emerge where complexes of lower operational principles form open systems, matrixes that allow for the infusion of new principles. Polanyi writes,

> The fluctuation which leads to the establish-
> ment of an open system does not vanish after
> the event....The atomic configuration which
> ignited a flame keeps renewing itself within
> the flame. It is a fundamental property of

> open systems...that they stabilize an improb-
> able event which serves to elicit them....
> The first beginning of life must have likewise
> stabilized the highly improbable fluctua-
> tion of inanimate matter which initiated
> life.[49]

Emergence is the result of "random" conditions that permit the activation of principles not to be found previous to those conditions. In Polanyi's words, emergence marks the advent of "an ordering principle capable of producing operational principles which the system had not previously possessed...."[50]

The question arises as to the exact origin of the new principles that will integrate the principles at a lower level in a new order. Polanyi clarifies his position by stating that,

> Random impacts can release the functions
> of an ordering principle and suitable
> physico-chemical conditions can sustain
> its continued operation; but the action
> which generates the embodiment of a novel
> ordering principle always lies in the
> principle itself.[51]

It would thus appear that ordering principles are not functions of the laws operating at the lower side of the boundary conditions. The ordering principles that oversee the functioning of living organisms, for example, are not determined by, although they are dependent upon, physio—chemical laws. Similarly, the higher principles

---

[49] Polanyi, _Personal Knowledge_, p. 384.

[50] Ibid., p. 399   [51] Ibid., p. 401.

of the mind would appear to be independent ordering principles that act in accord with, but are not derived from, biological principles. The concept of emergence, therefore, seems to be consistent with Polanyi's stratified entities and ontological levels.

However, closer examination of Polanyi's concept of emergence contradicts his arguments for a dualism and the contemplative freedom of the human mind from purely biological processes. The question of the ontological status of the ordering principles, including the ordering principles of the mind of man, has not yet been answered. Are ordering principles ontologically independent laws? Are they unique instances of physio-chemical mechanisms or are they as distinct as the meaning of the words on this page are from the ink which transcribes them? Is there a "far side" of reality that merges with the physical in man's thinking? Do mind and body constitute more than a rhetorical dualism?

Polanyi responds by stating that the entire course of evolution is completely corporeally-grounded. He asserts that anthropogenesis is entirely "a continuous proliferation of germ plasma, from unicellular origins to the germ plasm of the human couple of whom the man in question is born....This entire evolutionary achievement can be localized within a circumscribed material system."[52]

---

[52]Polanyi, Personal Knowledge, p. 386.

The entire system of evolution, from the beginning of
life through the rise of sentient and conscious life
forms to the rise of the mind of man, is, fundamentally,
the result of a series of complex material conditions
that permitted new ordering principles to emerge. These
ordering principles are simply physio-chemical laws
that become evident only in complex physio-chemical
fields.

There are no ontological entities transcending the
laws of physics and chemistry that insert themselves into physio-chemical mechanisms. Polanyi had begun his
inquiry into the human mind with the premise that physio-
chemical properties could neither be ascribed sentience
nor consciousness and that higher ordering principles had
to be assumed to act in living organisms. His conclusion,
however, rests upon the belief that sentience and consciousness can be ascribed to ordering principles that
are themselves merely physical and chemical. He provides
us with no explanation of why what he calls higher ordering principles--which are themselves but a subclass of
physio-chemical laws--should be considered capable of
giving rise to sensitivity, consciousness and, ultimately, mind.

Previously Polanyi's arguments seemed to indicate
that ordering principles inserted themselves into
material conditions where such conditions permitted.

This situation can be compared to a window which permits light to enter a room but does not itself create the light. It is now evident that ordering principles emerge from physical conditions in a similar fashion to a fruit that emerges from a growing tree. The point here is that there is no dualism. There is no life principle that _enters_ matter; there is no mind that _enters_ neurological mechanisms. Life is an extension of physio-chemical laws and man's mind is but one instance of further refinement. Polanyi's _dualism_ dissolves as the ontological status of the mind is found to be a rhetorical construct. No satisfactory explanation has been offered for the origin or ontological foundations of the contemplative demands of a mind that integrates ideas according to such requirements.

*   *   *

Although Polanyi's accounts of the ontological status of the mind and the mind/body relation are self-contradictory, he does provide a consistent model of mental acts when we simply accept the presence of contemplative principles. The notions of subsidiary/focal awareness and indwelling cannot be used to formulate ontological structures but rather functional relations. It is permissible to set aside Polanyi's arguments for a mind/body dualism and to consider consciousness from the

view that it is bipolar and intentional as would be
consistent with his general theory of the subsidiary/
focal nature of knowledge.

It is important to note that once the question of
the origin and ontological status of the mind are set
aside, the fundamental teleological organizational-principles
that would permit the contemplation of ideas according to
content is obscured. In short, the impetus for, the
underlying functions of, the quest for knowledge of universal standing is left unjustified. Polanyi explains
that physics and chemistry cannot explain the existence
of teleological principles and specifically cannot understand contemplative aims and commitments. Consequently,
his failure to consistently develop his notion of the
integrity of the mind with its unique organizational
principles dissolves his arguments for purely contemplative aims. Polanyi states, "Though rooted in the body,
the mind is free in its actions."[53] His explanation for
the freedom of the mind, however, is inadequate. How can
the mind be free to pursue teleological aims that are unavailable to the neurological processes of the brain when
the brain is the absolute generator and organizer of
human thought? If the whole course of emergence is completely "materially circumscribed," the mind could not

---

[53]Polanyi, American Psychologist, p. 40.

introduce goals and standards that had not first been
established as electro-chemical processes in the brain.
Thus restrained, let us briefly examine Polanyi's notion
of the rise of articulate thought.

Whatever the ontological status and the organizational
principle of the human mind, Polanyi claims that the activity of constructing complex systems of articulate thought
results from tacit intellectual capabilities. He explains
that "the towering superiority of man over animals is due,
paradoxically, to an almost imperceptible advantage in
his original, inarticulate facilities."[54] The slight inarticulate advantage enjoyed by humans enable man to
create and integrate conceptual symbols that are capable
of numerous and varied applications. Polanyi states,

> By responding to people who talk to it,
> the child soon begins to understand speech
> and to speak itself. By this one single
> trick in which it surpasses the animal,
> the child acquires the capacity for sus-
> tained thought and enters on the whole cul-
> tural heritage of its ancestors.[55]

Such a contention may appear to be contradictory to
Polanyi's emphasis on the tacit aspects of knowledge.
However, as we stated in Chapter II, "all knowing is
either tacit or tacitly rooted." Verbal knowledge is
tacitly rooted. Language, according to Polanyi, is an

---

[54]Polanyi, _Personal Knowledge_, p. 69.

[55]Ibid.

intellectual tool used for symbolic representation. As a tool, it can only be used properly when one has a subsidiary notion of its uses, aims, characteristics, and rules. Language can neither be developed nor sustained where subsidiary integrative powers are lacking. The implications of this perspective are clear. Polanyi explains,

> If, as it would seem, the meaning of all our utterances is determined to an important extent by a skillful act of our own--the act of knowing--then the acceptance of any of our own utterances as true involves our approval of our own skill. To affirm anything implies, then, to this extent an appraisal of our own art of knowing, and the establishment of truth becomes decisively dependent on a set of personal criteria of our own which cannot be formally defined. If everywhere it is the inarticulate which has the last word, unspoken and yet decisive, then a corresponding abridgement of the status of spoken truth itself is inevitable.[56],*

Our present task will be to clarify the relations Polanyi sees between inarticulate integrative powers and articulate systems of thought. We shall see that Polanyi maintains that contemplative structures and aims emerge in the course of the evolution of culture. We must bear in mind that their emergence does not result from the insertion of higher contemplative principles. Our inquiry begins with the recognition of a de facto

---

[56]Polanyi, _Personal Knowledge_, pp. 70-71.

*It should be noted that the opening sections of Chapter II dealt with Polanyi's contention that mathematical procedures rest upon tacit grounds of assent. The

increase in knowledge and skill, both culturally and individually. With such limitations in mind, let us briefly approach Polanyi's concept of the development of articulate systems of thought.

We may begin with a brief analysis of three different types of inarticulate intelligence distinguished by Polanyi. Polanyi develops his concept of inarticulate intellectual powers by referring to classes of intellectual functions performed by animals. This perspective is consistent with his belief that human intelligence and animal intelligence share the same subsidiary/focal character. In order for human beings and animals to attend to a given stimulus, they must do so from within a subsidiary context; they must come to dwell heuristically in the phenomena that they perceive. Polanyi states, "Fundamental novelty can be discovered only by the same tacit powers rats use in learning a maze."[57]

Such notions are also consistent with Polanyi's concept of emergence. In this case the operational principles functioning in animal intelligence leave an open border that can permit the emergence of formal articulate structures. Human beings, Polanyi claims, having a slightly

---

last section of this chapter, therefore, will focus upon non-mathematical language to demonstrate the tacit-rootedness of all language and further, describe the relation between articulate structures and such inarticulate underpinnings.

[57]Polanyi, Study of Man, p. 18.

more highly developed level of inarticulate intelligence can provide sufficient conditions for articulate thought to develop.

However, we must remember Polanyi fails to clearly indicate the source of nature of the higher organizational principles of human intelligence. Consequently, he is unable to distinguish factors in human intelligence that may be discontinuous with the intellectual functions of animals. For example, Polanyi stresses the importance of the _ideal_ of universal intent, an ideal of sufficient force to override survival instincts and subjective desires and goals. Science would cease to exist for Polanyi without such an ideal; science is shaped by it. Yet, Polanyi cannot account for its existence, and he fails to differentiate its role in superintending human consciousness as it engages in scientific inquiry.

The first type of inarticulate intelligence is demonstrated in acts of what Polanyi calls "trick learning." A rat in a Skinner box learning to depress a lever to release a pellet of food provides a clear example. Polanyi claims that such an act of learning includes a teleological element, a means-end relation, that is precluded from logical positivistic-type inquiries. He states that the amplification of a rat's behavior is elicited by "providing an object which it can use as a tool, and it consists in discovering and practicing the proper use of this tool."[58] Even the most simple forms of learning involve bringing a subsidiary, i.e. a lever, to bear on a focus, i.e. food pellets. A tacit, purposive framework thus provides the context for the observed behavior.

The second type of inarticulate act of intelligence distinguished by Polanyi he calls "sign learning." This type of intelligent activity is evidenced (and frequently misinterpreted) in Pavlov's famous classical conditioning experiments. Unlike Pavlov who claims that conditioned stimuli act as if they were unconditioned stimuli, Polanyi notes that "the 'conditioned' response differs from the original 'unconditioned response,' in the same way in which the anticipation of an event differs from the effect of the event itself."[59,60] The teleological element, the tacit integrational process, here presents itself in the subject's underlying capacity to recognize a sign as foretelling an event. A similar experiment where rats were systematically introduced to the task of finding food behind one of two doors by identifying a specific sign indicating the presence of food demonstrates, according to Polanyi, not a simple act of conditioning, but rather the presence of a complex of subsidiary factors that play into animal behavior. He states that the simple act of

---

[58]Polanyi, _Personal Knowledge_, p. 72

[59]Ibid.

[60]Polanyi's theory of indwelling provides a logical

perceptional discrimination performed by the rats shows

the animals'

> capacity to be intrigued by a situation, to
> pursue consistently the intimation of a
> hidden possibility for bringing it under
> control, and to discover in the pursuit of
> this aim an orderly context concealed be-
> hind its puzzling appearances. The essen-
> tial features of problem-solving are thus
> apparent even at this primitive level.[61]

The third type of inarticulate intelligence dis-

tinguished by Polanyi is evidenced in what he calls acts

of "latent learning." Whereas trick learning is primari-

ly motile and sign-learning is primarily perceptual,

latent learning is fundamentally heuristic; it estab-

lishes an understanding of the workings of an object--

the principles of an object's operations. Latent learn-

ing involves an animal's ability to subsidiarily reinte-

grate the information derived from a given experience so

as to apply it to new and unexpected situations; in es-

sence, an act of latent learning establishes a subsidiary

heuristic framework through which an animal may antici-

pate the behavior of an object under various yet to be

experienced conditions. Such heuristic achievements,

Polanyi maintains, are found in rats, which having

---

framework for making such a distinction. A stimulus
used to predict a given occurrence is itself a subsidiary
from which to focus upon the expected event. Stimuli
thus acquire a semantic aspect whereas originally they
had none. Past events thus create a dwelling place which
can be used to attend to new focuses.

[61]Polanyi, Personal Knowledge, p. 73.

mastered a given maze, can exercise ingenuity in over-coming obstacles placed within that maze.[62] Along with this increased possibility of discovering new relation-ships comes the increased possibility that inadequate or inaccurate assumptions can breed incorrect expectations. Polanyi writes, "Thus, the very rise of inferential power brings with it the conjoint capacity for inferential error."[63]

The operations by which the initial acts of trick, sign, and latent learning take place are irreducible. Polanyi classifies them as a "plunging-in" in which the animal responds to a stimulus with his own activity using the vague and perhaps confused conditions as clues to the meaning of an object or event. Such subsidiary in-corporation of stimuli, their assimilation as aspects of the animal's intellectual dwelling place, does not pro-ceed by way of formal operations; data are assimilated as informal indicators which can be used to reveal some as yet hidden aspect of reality. Further, this process is irreversible; retracing the conclusions back to sup-posed assumptions would not reveal how initial indicators took on the meaning they were assigned. Irreversibility is a general feature of all heuristic acts that add to

---

[62]Polanyi, Personal Knowledge, p. 74.

[63]Ibid.

existing knowledge. At the most sophisticated end of
intellectual activity, Einstein's theory of relativity,
for example, could be traced back to its premises in a
logical sequence, but such a procedure would never reveal the unformalizable heuristics that Einstein employed
to originally assign the unique significance he did for
the data he had available.[*]

Subsequent uses of knowledge discovered through
"plunging" into a maze of stimuli, however, can be traced
and understood in terms of fixed relations and expectations. "Routine" operations, as they might be called,
are both reducible and reversible; they follow the pattern laid by the initial heuristic act. Polanyi concludes that even at the primitive level of the examples
given there exist "two kinds of intelligence: one achiev-
ing innovations, irreversibly, the other operating a
fixed framework of knowledge, reversibly."[64]

Thus Polanyi provides a basis for exploring the relation between inarticulate and articulate thought. Articulation proceeds by virtue of heuristic acts which,
being irreducible, informal and inarticulate, evade the
probing inquiries of critical reflection. The tacit roots
of articulate systems of thought are intellectually

---

[*]Chapter II includes a more detailed discussion of
the notion of irreversibility.

[64]Polanyi, _Personal Knowledge_, p. 76.

(formally and explicitly) invisible.

Polanyi explains the generation of articulate thoughts in the following manner. He states that inarticulate powers of intelligence allow human beings to use an extended number of subsidiary factors in their attempt to gain intellectual mastery over a given focus. In the same way a rat becomes intrigued with and consistently pursues the structure of a maze, the human being can become intrigued with and pursue the structure of language. Through acts of sustained attention, Polanyi claims that human beings begin to assimilate the content and form of language. The process of assimilation is one in which the learner attempts to integrate various words and grammars in a subsidiary fashion, that is, as he imagines they function in the as yet undiscovered laws of language. Polanyi states, "My view is that the use of language is a tacit performance; the meaning of language arises, as many other kinds of meaning do, in tacitly integrating hitherto meaningless acts [in this case spoken words] into a bearing on a focus that thereby becomes their meaning."[65]

Polanyi refers to the activity of generating new integrations, of weaving sets of subsidiaries, of applying unique subsidiaries arising in a particular situation,

---

[65]Grene, Knowing and Being, p. 196.

as acts of imagination. Polanyi explains,

> The casting forward of an intention is an
> act of imagination. It is only the imagi-
> nation that can direct our attention to a
> target that is as yet unsupported by sub-
> sidiaries.[66]

The "casting forward" of the imagination is identical with Polanyi's earlier term "plunging in." In this
case, the individual has enough subsidiary information
to make attempts to formulate a new integration of a
focus. The phrase "a target that is as yet unsupported
by subsidiaries" refers to a focus which, because of the
new subsidiary integration required, is yet vague and un-
clear. Polanyi does not believe that such acts of
imagination are either haphazard or formal. The imagin-
action is guided by a heuristic vision of reality, a
vision replete with implication. The imagination extends
the previous parameters of knowledge in accordance with
an informal subsidiary understanding of that knowledge.

The thrusting forth of the imagination is thus
preconceptual. Articulate structures rise on the new
ground created in the imaginative extension of knowledge.
The process is often experienced in the common occurrence of forgetting a word. We have all had the experience where we know what we mean or want to say and yet
cannot find the words to communicate our initial meaning.

---

[66]Polanyi, American Psychologist, p. 40.

The imagination guides us in our search for the lost word. We often employ various techniques as subsidiaries to the vague, yet anticipated, focus. The imagination is the cutting edge of thought that pierces the unknown. It extends the subsidiary foundations of one's fundamental vision of reality. Polanyi notes that it is possible for us to speak only because

> we feel that many thousands of words are available for our novel purposes, and we can trust the powers of our imagination, bent upon this purpose, to evoke from these available resources, the implementation of our purpose.[67]

The rules of grammar are also learned with an initial thrust of imaginative activity. We subsidiarily rely upon the rules of grammar to make sense of another person's speech or a written text. These rules are recreated imaginatively by each person learning to apply them in listening to or generating language. Polanyi concludes that the "striving imagination has the power to implement its aims by the subsidiary practice of the ingenious rules of which the subject remains focally ignorant."[68]

It is most important, however, to understand that the imagination is accompanied in its efforts by what Polanyi calls the "intuition." The imagination is

---

[67]Polanyi, American Psychologist, p. 40.

[68]Grene, Knowing and Being, p. 200.

guided by a general integration of subsidiaries;
the anticipations and expectations that the imagination
thrusts are framed for the imagination intuitively. It
is the intuition that anticipates the need for an existence of a needed work; it is the intuition that confers
judgment of another person's meaning; it is the
intuition that selects the grammatical structures suitable for melding imaginative thrusts into articulate
ideas. The imagination and intuition interact continually in the processes of making sense of and generating
language. Polanyi states,

> And it is the intuition that forms there
> our surmises and which eventually selects
> from the material mobilized by the imagina-
> tion the relevant pieces of evidence and
> integrates them into the solution of the
> problem.[69]

The imagination may generally be understood to be
an active, generative, inarticulate (in the case of man,
prearticulate) intelligence, while the intuition may be
said to be a heuristic power of inarticulate (prearticulate) judgment. We will consider these informal intellectual capacities in greater detail in Chapter IV where
we extend the whole process of articulation into the process of conceptualizing scientific theory. It is appropriate at this time, however, to call attention to the

---

[69]Polanyi, American Psychologist, p. 42.

idea that articulate structures and processes geometric-
ally expand the informal capacities of the imagination
and intuition. Articulate structures permit the storage
of vast amounts of information including the intellectual heritage of past generations. Further, articulate
and/or formal intellectual processes enable us to isolate
our intuitive expectations and integrations and examine
their consistency and cohesiveness. Polanyi explains
that speech enables man to systematize informal and to
conduct inferential processes without reference to particular measures or entities. Both the articulate structures and the conclusions of formal processes provide
vast regions of implication to challenge the integrative-
ness of the intuition and the effusiveness of the imagination. Articulate entities permit the human "mind" to
extend itself beyond the limitations of the time, space
and isolation of one's own particular experience.

Polanyi further maintains that there are operational
principles of language that provide a dynamic context
for the workings of the imagination and intuition as
they operate subsidiarily in the generation and reception of symbols. (His concept of language includes all
types of articulate structures including mathematics.)
The first principle governs the process of symbolic representation and the second oversees the operation of
the resultant symbols as they act in the process of

conceptualization.

1) The first operational principle has several aspects: the law of poverty and the law of grammar. The law of poverty states that language must have a sufficiently limited number of symbols so as to allow for their meaning to develop through repeated usage. One learns the meaning of a word by assimilating its subsidiary function as one attempts to attend to a meaning one wishes to express. Lack of subsidiary knowledge of a word--knowledge of its appropriate application--renders it meaningless.

The law of grammar allows for the consistent yet highly flexible use of symbols by providing fixed patterns of combinations. Polanyi explains, "Only grammatically ordered word clusters can say with a limited vocabulary the immense variety of things that are apposite to the range of known experience."[70]

Underlying these laws are two more fundamental requirements. The first, called the law of iteration, states that if a symbol is to be useful, it must be identifiable. A word undistinguishable from other words is useless. The second, called the law of consistency, is far more significant. It states that in spite of the peculiarities of individual situations, symbols must be applied homologically. Thus, the use of a word implies

---

[70]Polanyi, _Personal Knowledge_, p. 79.

that one has made personal judgments about the features
of that which it symbolizes, the identification of
those features in a particular situation and the significance of variation of features within that situation.
Together these laws

> imply that every time we use a word for
> denoting something, we perform and accredit
> our performance of an act of generalization
> and that, correspondingly, the use of such
> a word is taken to designate a class to which
> we attribute a substantial character.[71]

Further, the generalizations embodied in our language not only help us cohesively integrate present phenomena, they also create patterns of integration that we
expect may be used in the future. Polanyi maintains that
so long as we feel satisfied with the classifications in
our language, we commit ourselves to the vision of the
world incorporated therein.

Polanyi believes that the imagination and intuition are informally active in developing categories.
Consequently, the categories themselves are not circumscribed by the term that symbolizes them. The classification of man, for example, transcends the definition
of the term "man." Judgments made about a particular
instance of "man" need not depend upon the explicit definition but rather upon the subsidiary integrations
that are incompletely described in explicit form.

---

[71]Polanyi, Personal Knowledge, p. 80.

Polanyi states that in speaking of "man" we are not attending to any kind of man, but relying on our subsidiary awareness of individual men, for attending to their joint meaning.[72] The identification of an object of a given class proceeds at a largely inarticulate level. "Denotation, then, is an art, and whatever we say about things assumes our endorsement of our own skill in practicing this art."[73]

2) The second operational principle of language consists in its ability to represent various objects and relations in a manageable form. "Language can assist thought only to the extent to which its symbols can be reproduced, stored up, transported, re-arranged, and thus more easily pondered, than the things which they denote."[74] Polanyi refers to this aspect of language as the law of manageability.

The fact that objects and events can be represented in symbols enables us not only to denote such phenomena but also to explore and examine their characteristics. Symbolic representations can reduce the scale of events occurring over vast stretches of time and space; minute flashes can be subdivided into detailed phases.

---

[72]Grene, Knowing and Being, p. 149.

[73]Polanyi, Personal Knowledge, p. 81.

[74]Ibid.

Further, the tractability of symbolic representations permits us to reorganize given phenomena into new sets revealing new relationships and patterns. Graphs and equations, for example, can be drawn from lists of measures. Polanyi concludes, "Articulation pictures the essentials of a situation on a reduced scale, which lends itself more easily to imaginative manipulations than the ungainly original...."[75]

The operational principles of language reveal that language itself was created and is sutained by inarticulate powers of intelligence. "To speak is to contrive signs, to observe their fitness, and to interpret their alternative relations."[76] To speak implies that one dwells within language, that one has a subsidiary understanding of its structure and function. Language is, in the last analysis, a tool for shaping and expanding experiences in accordance with the vision underlying language itself and infusing it with meaning.

In this chapter we have examined Polanyi's concept of indwelling to clarify the inarticulate foundations of human thought in experience. Contradictions were found in Polanyi's ontology of the human mind, and the teleological tenets of human thinking were therefore found

---

[75] Polanyi, Personal Knowledge, p. 85.

[76] Ibid., p. 82.

to be obscure. However, Polanyi's account of the rise and structure of human thought was consistent and provided a cohesive schematic for understanding the interplay of articulate and inarticulate intellectual powers. The following chapter will further detail the relation between the Polanyi concepts of the imaginative and intuitive aspects of inarticulate thinking and the process of formulating explicit scientific theories.

Chapter IV

THE PROCESS OF DISCOVERY

The three previous chapters have explored Polanyi's notions of the ontological referents of science, the unspecifiable foundations of scientific inquiry, the tacit triad structure of human knowledge and the processes that give rise to formal, articulate structures. The purpose of this chapter is to integrate these ideas and apply them to Polanyi's account of the process of scientific discovery.

As one might expect, Polanyi asserts that the process of scientific discovery does not begin with the formation of hypotheses based upon observation alone; it is a highly complex process beginning and ending with the personal heuristic powers. Polanyi explains that scientific discovery begins with the personal[1] recognition of a problem that requires active pursuit. He states,

> All true scientific research starts with
> hitting on a deep and promising problem,
> and this is half the discovery. Is a
> problem a hypothesis? It is something
> much vaguer. Besides supposing the dis-
> covery of a problem were replaced by the
> setting up of a hypothesis, such a

---

[1]Michael Polanyi, "Problem Solving," The British Journal for the Philosophy of Science VII, 30 (August 1957):92.

> hypothesis would have to be either one
> formulated at random or so chosen that
> it has a fair initial chance of being
> true.[2]

Thus it is clear that the selection of problems suitable for inquiry requires the personal assessment of the unknown, of the potentialities of aspects of reality that have hitherto remained hidden. The selection of a valuable problem, however, provides more than a mere subject for inquiry; it is a heuristic anticipation, a vision of a coherence that, for reasons perhaps quite unspecifiable, the researcher believes to exist. "It's an engrossing possession of incipient knowledge which passionately strives to validate itself. Such is the heuristic power of a problem."[3]

Yet further, a problem can be understood as a surmise, an accessment of a hidden coherence that acts as a focus for inquiry. Consequently, empirical data are not viewed in and of themselves but as clues to or indicators of the laws which they only partially embody. Polanyi claims, as noted in Chapter I, that reality is rational in the sense that it is lawful and orderly; our thinking may be said to be rational to the extent it mirrors in itself the orderliness and lawfulness of

---

[2]Marjorie Grene, _Knowing and Being_ (Berkeley: University of California Press, 1974), p. 118.

[3]Ibid., p. 132.

ontological reality. The rational coherence that gener-
ates the characteristics of the phenomenal field observed
by a researcher occupies his central focus; Polanyi
concludes, "The knowledge of a problem is, therefore,
like the knowing of unspecifiables, a knowing more than
you can tell."[4] One senses a rational coherence in the
formulation of a problem but cannot specifically identify its nature or character. If one could be more specific, the problem would no longer be a problem.

This paradoxical knowing and yet not knowing a
rational coherence in the form of a problem is resolved
when we recognize the tacit triadic nature of all knowledge.
Perhaps the simplest explanation of the tacit triadic
structure of problems generating scientific inquiry can
be found in the act of perception. Polanyi maintains that
perception is a personal activity in which an individual
uses numerous subsidiary factors to attend to an object
which he has yet to clearly distinguish. Most of the
time the integrations of subsidiary factors is routine
and requires little conscious effort but subsidiaries
are always actively integrated to produce a clear focus.
Polanyi states that "the efforts of perception are
evoked by scattered features of raw experience suggest-
ing the presence of a hidden pattern which will make

---

[4]Grene, _Knowing and Being_, p. 141.

sense of the experience."[5] The raw sensory impressions we receive act as subsidiaries, as indicators of a coherence of which they themselves are only parts. The object perceived is not perceived as a distinct object until the subsidiaries of past experience (eye musculature, peripheral visual stimuli, etc.)[6] are integrated sufficiently to give it clarity. Polanyi concludes that this aspect of "perception is performed by straining our attention toward a problematic centre, while relying on hidden clues which are eventually embodied in the appearance of the object recognized by perception."[7]

Polanyi relates the subsidiary use of particulars in attending to an object being perceived to the subsidiary use of empirical data in attending to rationally intelligible coherences or, simply, natural laws. It is undoubtedly so that the latter process requires sophisticated and highly systematic articulate structures and standards. However, it is essential to note, as explained in Chapter II, that even the most formal and explicit procedures of science are personally accredited, however, unconsciously, and utilized with personal judgment and unformalized subsidiary facility; they are

---

[5]Grene, _Knowing and Being_, p. 117.

[6]Ibid., p. 115.

[7]Ibid., p. 118.

applied to discerning a coherence indistinguishable to
less rigorous subsidiary structures.

Polanyi offers the following example. He states
that it is a simple matter to perceive shapes such as
balls or eggs, but that it would be far more complex to
perceive such shapes if, instead of being presented with
an aggregate of sensory stimuli forming the surface of
the object, we were given the logical equivalent of these
points in the form of spatial coordinate values. Polanyi
concludes that "the perception of the egg from the list
of coordinate values would, in fact, be a feat rather
similar in nature and measure of the intellectual achievement to the discovery of the Copernican system."[8] Copernicus was presented with a multitude of empirical
phenomena from which he was to select and integrate those
indicative of a coherence that was beyond his present
knowledge. His subsidiary dwelling place was far more
complex than that used in ordinary perception but was
similar to perception in the use of subsidiaries to attend to a problematic center, a vaguely anticipated
coherence. Polanyi concludes,

> We may say, therefore, that the capacity of
> scientists to guess the presence of shapes
> as tokens of reality differs from the capac-
> ity of our ordinary perception, only by the
> fact that it can integrate shapes presented

---

[8]Michael Polanyi, Science, Faith and Society
(London: Oxford University Press, 1946), p. 24.

162

to it in terms which the perception of
ordinary people cannot readily handle.[9]

It is easy to misinterpret Polanyi's point.
Firstly, he does not mean that the only difference,
per se, between perception and scientific inquiry lies
in the complexity of the coherence considered; numerous
sociological, logical and procedural factors enter into
the conduct of inquiry that have only diffuse equivalents
in the act of perception. Polanyi's intention is, rather,
to highlight the similarity in general structure between
scientific inquiry and perception. Secondly, Polanyi's
use of the word shape is not limited to spatial entities;
he uses the word to refer to all varieties of empirical
phenomenal patterns believed to be indicative of rational
coherence.

We can conceive of the problematic core of the
process of discovery within the context of the tacit
knowing triad. In this case, the research scientist (A)
uses empirical data (B)[10] to attend to a rational coherence (C). In its problematic stage C is but a vaguely
anticipated entity or pattern. Problems may thus be
generally defined as focal entities having an unspecifiable or simply an empty center. If the problem is well
chosen and the subsidiaries sufficient, a gradual

[9]Polanyi, Science, Faith and Society, p. 24.

[10]Empirical data constitute only one part of the
research scientist's dwelling place; his knowledge of

clarification of C should occur as integrative efforts
continue. Polanyi explains that

> A problem designates a gap within a
> constellation of clues pointing towards
> something unknown. If we hold a problem
> to be a good one, we also imply that this
> unknown can yet be discovered by our own
> efforts and that this would be worth
> these efforts.[11]

Polanyi quotes a response by Polya to a student's
inquiry as to how he could become a great researcher.
Polya said, "Look at the unknown! Look at the end.
Remember your aim. Do not lose sight of what is required. Keep in mind what you are looking for. Look
at the unknown. Look at the conclusion.'"[12] Polya's
advice is in keeping with the tacit triadic character of
problems. Polya is, in essence, advising his student to
review the empirical data available not as objects unto
themselves but as indicators to and parts of an unknown
coherence. It is only in relation to the rational coherence being sought that empirical particulars derive
their function and meaning.

It is, of course, possible to use formal procedures
to integrate data to arrive at a conclusion. However,
procedures are not randomly applied but purposively

---

particular theories, techniques, standards, and most importantly, his own vision of the world also act as subsidiary factors.

[11]Grene, _Knowing and Being_, p. 171.

[12]Polanyi, "Problem Solving," p. 98.

selected; they are used to put the data in a form that the researcher can use to evaluate the phenomena. The initiation of discovery therefore begins with a surmise of what one expects to be a fruitful line of inquiry (a problem); the process of discovery progresses along a path guided by such heuristic expectations.

The process of discovery incorporates formal articulate structure as tools guided by the researcher's personal vision. The scientist's vision is here composed of a weaving of informal and formal items into a vague outline of the unknown. The process of discovery implies, according to Polanyi, that a researcher cross a logical gap between the known and the unknown, between the existing body of knowledge and the hidden aspects of reality he believes to be present. Polanyi explains that "the pioneer mind which reaches its own distinctive conclusions by crossing a logical gap deviates from the commonly accepted process of reasoning, to achieve surprising results."[13]

I recently attended a lecture on the philosophy of scientific discovery by the Novel prize winning immunologist, Professor David Blumberg.[14] In that lecture

---

[13]Michael Polanyi, Personal Knowledge: Towards a Post-Critical Philosophy (New York: Harper & Row, Publishers, 1964), p. 123.

[14]The Dankin Memorial Lecture given at Adelphi University, March 31, 1981.

he described how he believed scientific discoveries were
generated. He presented the audience with a drawing
composed of a number of dots (see Fig. 1). The dots
represented a sampling of empirical phenomena that are
available to researchers at the beginning of their research. Subsequent images containing increasing numbers
of dots were presented, and Dr. Blumberg began to connect
the dots in recognizable patterns (see Figs. 2 and 3).
The increasing number of dots depicted new evidence available to researchers in the course of their inquiries.
The connecting of the dots represented the step by step
process of using the available data to systematically arrive at a coherent intellectual analysis of the phenomena,
or simply a scientific discovery.

These Figures are approximations of those
used by Dr. Blumberg

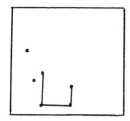 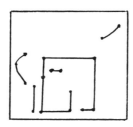 

Fig. 1   Fig. 2   Fig. 3

The central problem with Dr. Blumberg's analysis lies in his lack of recognition of the fact that one could not connect the dots into a recognizable pattern unless one had a general expectation of what would constitute an identifiable pattern. It is quite possible to derive any number of distinct relations between a maze of dots (see Fig. 4). Only that series of relations the researcher believes to be indicative of ontological order is selected as valuable to science. Each and every choice a scientist makes from the selection of this problem to the methods he uses, from judgments he makes to the direction he takes in the process of the inquiry are guided, fundamentally, by his own vision of reality, a vision he shares, in part, with his contemporaries.

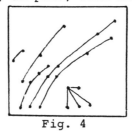

Fig. 4

This figure illustrates one of an infinite variety of patterns that can be formed from the same complex of dots said to be indicators of a single pattern. "Never yet has a definite rule been laid down by which any particular mathematical function can be recognized, among the infinite number of those offering themselves for choice, as the one which expresses a natural law."[15]

---

[15]Polanyi, Science, Faith and Society, p. 21.

Further, the act of discovery may reveal new and unprecedented conceptions of reality making the recognition of coherent patterns more complex. Once a pattern is recognized, it appears that all the empirical data considered are logically integrated. The fact of the matter remains that the initial discernment of the coherence was not itself a formal, articulate process but a tacit act of integration guided by the personal vision of the scientist. The specifics of the act of integrating subsidiaries in the process of scientific discovery will be analyzed in detail later in this chapter. What is significant for us now is to recognize that scientific discovery begins with a problem that opens a logical gap incapable of being systematically spanned. Polanyi concludes that scientific discovery is an irreversible process in that the *original* act of arriving at a coherence cannot be traced back from its conclusions to its premises in a strict, logical manner. The selections a scientist makes with regard to problems, judgments, and direction for inquiry are tacitly grounded even though the conclusion may be logically reducible. Polanyi states that,

> true discovery is not a strictly logical
> performance, and accordingly, we may
> describe the obstacle to be overcome in
> solving a problem as a "logical gap,"
> and speak of the width of the logical
> gap as the measure of the ingenuity required for solving the problem.[16]

---

[16]Polanyi, *Personal Knowledge*, p. 123.

Accordingly, we should expect that the capacity to cross the logical gap necessary to arrive at a scientific discovery will require personal gifts of creativity and originality on the part of the scientist. Polanyi necessarily emphasizes the role of such personal intellectual gifts in the process of scientific discovery. He states, "genius makes contact with reality on an exceptionally wide range; seeing problems and reaching out to hidden possibilities for solving them, far beyond the anticipating powers of current conceptions."[17] The nature of this notion of genius will be found in the imaginative and intuitive aspects of the process of scientific discovery. Discovery is not accomplished systematically; it is filled with continuous possibilities for misdirection and error; it requires exceptional intellectual gifts to subsidiarily transform the known into valuable clues for revealing the unknown.

Before we direct ourselves to Polanyi's account of the mechanisms operating in scientific discovery, we must first answer two further questions about scientific problems. Firstly, Polanyi indicates that there are good problems and poor problems; we must ask what constitutes a good problem. Secondly, if there are such things as good problems, how can we understand how they are found?

---

[17] Polanyi, _Personal Knowledge_, p. 124.

The first issue is answered by Polanyi's notion of the ontological referent of science. This ontological referent does not indicate that science is concerned solely with clearly identifiable empirical phenomena. Rather, Polanyi believes that the operational principles creating the phenomena we perceive are the proper subject of science (see Chapter I). Thus, the reality that science approaches is not phenomenally complete and fixed but partial and generative with numerous aspects that have yet to be revealed. Polanyi writes that "reality is something that attracts our attention by clues which harass and beguile our minds into getting ever closer to it, and which, since it owes this attractive power to its independent existence, can always manifest itself in still unexpected ways."[18] Consequently, a good scientific problem focuses upon an ontological coherence so that the solution to the problem in the form of the recognition of such a coherence is capable of establishing a wide variety of new contacts with reality. A good problem is one by which subsidiaries can be integrated into a coherence while a poor problem diffuses subsidiaries leaving only a distorted and hence heurstically inadequate context (see Chapter III).

The question of how to arrive at good problems,

---

[18]Grene, _Knowing and Being_, pp. 119-120.

other than as a random function, is a far more complex matter. Researchers, in the course of their scientific education, not only learn specific theories but subsidiarily assimilate a mode of approaching and a general understanding of the world. The assimilation of the scientific vision of reality is accomplished through learning how it functions rather than through an explicit analysis only of its scientific theory and deductive implications. In the process of scientific training the scientist fiduciarily commits himself to the vision of the world undergirding his study; to this extent he assumes an intellectual dwelling place from which to approach his research. Of course, the particularities of an individual's experience and capacities personally shape his dwelling place, his heuristic vision. However, Polanyi indicates that "even the highest degree of intuitive originality can operate only by relying to a considerable extent on the hitherto accepted interpretative framework of science."[19]

A scientist's intellectual foundations constitute a significant heuristic facility to the extent that the presuppositions embodied therein function to integrate phenomena in the same way that the operational principles functioning in a given ontological field determine the

_____

[19]Grene, Knowing and Being, p. 119.

characteristics of the phenomena perceived (see Chapter
III). Thus, a scientist can arrive at significant and
fruitful problems to the degree that his heuristic vision
interprets phenomena in accordance with the ontological
operational principles governing the action of the phenomena. To the extent that reality is understood to consist
in that "which is expected to reveal itself indetermin-
ately in the future,"[20] heuristic visions paralleling the
orderly structure of reality should be able to anticipate
significant problematic areas requiring the further refinement of the initial vision.

Polanyi explains,

> This vision, the vision of a hidden
> reality, which guides a scientist in his
> quest, is a dynamic force. At the end
> of the quest the vision is becalmed in
> the contemplation of the reality revealed
> by a discovery; but the vision is renewed
> and becomes dynamic again in other scien-
> tists and guides them to new discoveries.[21]

Modern scientists create their intellectual dwelling place with reference to centuries of scientific
traditions and discoveries. Scientists not only apply
theories in systematic and deductive fashions (recognizing, of course, the continuous need for personal judgment
and commitment) but apply them to novel situations in

[20]Polanyi, Science, Faith and Society, p. 10.

[21]Michael Polanyi, "The Creative Imagination,"
Tri-Quarterly 8 (Winter 1967):113.

creative and original ways; scientists use theories, in such cases, as subsidiary clues to the hidden reality they seek. In this way, theories act as vague premises for formulating expectations. Polanyi states that such premises

> ...indicate to scientists the kind of
> questions which seem reasonable and inter-
> esting to explore, the kind of conceptions
> and relations that should be upheld as
> possible, even when some evidence seems to
> contradict them, or that on the contrary,
> should be rejected as unlikely, even though
> there was evidence which would favour them.[22]

There are no specific guidelines as to the types of empirical clues inquiring scientists pursue. Polanyi indicates that some discoveries have resulted from focus-ing upon regularities in nature while others have begun by focusing upon unexpected irregular empirical data. Still other discoveries have resulted from seeming con-flicts between existing theories and yet others from the possibility of subsuming one within another. "In any case, such [categories of] clues can merely offer a possible occasion for discovery; they do not tell us how to make a discovery...."[23]

The question might arise as to why, if the heuristic

---

[22]Polanyi, _Science, Faith and Society_, p. 11.

[23]Grene, _Knowing and Being_, p. 202.

vision of the scientist is to some degree accurate to the ontological principles operating in a given field, specific, explicit inferences should not be forthcoming. The answer is quite simple: scientific discoveries are as complex and comprehensive as they need to be in order to respond to the problems they address. Scientific theories do not attempt to explain facets of reality that have not been necessary to research in order to answer the questions they have about the phenomena they consider. It would not have been necessary for the forerunner of atomic theory to predict the multitude of subatomic particles now believed to exist. The discovery of such particles arose only when atomic theory created an area of inquiry opening questions that necessitated such refinement.

The premises underlying theories are unspecifiable because they are both subsidiary to the formulation and implementation of theories and incomplete in the sense that they constantly open up new problematic areas. Thus, Polanyi succinctly concludes,

> Science is based on clues that have a
> bearing on reality
>
> These clues are not fully specifiable
>
> Nor is the process of integration which
> connects them fully definable
>
> And the future manifestations of the
> reality indicated by this coherence are

inexhaustible.[24]

The presuppositions constituting a scientist's
heuristic vision are unspecifiable and fiduciary (see
Chapter I). They are not subject to formal, systematic
interactions; phenomena inconsistent with their joint
implications need not refute them as a body (see Chapter
II). It is only when phenomenal inconsistencies can be
anticipated to form meaningful patterns that the original
premises that gave use to the inquiry will be transformed. The selection of problems and the eventual evaluation of the success of theoretical solutions are informally judged by the dynamic force of the scientist's
heuristic visions.

Polanyi notes, for example, that Copernican theory
was based on "a veritable jungle of ad hoc assumptions."[25]
He also points out that in spite of significant mechanical
problems and new conceptions about the distance of the
stars, Copernicus continued to believe that his system
revealed harmonies that he himself could express in a cohesive, coherent manner. Polanyi concludes that Copernicus

> did not stop to consider how many assump-
> tions he had to make in formulating his

[24]Polanyi, Tri-Quarterly, p. 116.

[25]Ibid., p. 112.

system, nor how many he ignored in doing
so. Since his vision showed him an outline of reality, he ignored all its com-
plications and unanswered questions.[26]

We must ask how, if the heuristic vision guiding
a scientist in the pursuit of discovery is so indeterminate and unsystematically formulated, it can be utilized by the scientist.[27] Once again, Polanyi reminds
us that scientific inquiry embodies the same general
structure as perception. He maintains that the integrative capacities manifest in the assimilation of disparate
sensory impressions in the act of perception is but a low
order of the integrative power of the pioneering scientist in the act of discovery. The argument is thus made
that the informal intellectual powers needed to perform
complex and problematic perceptions can be found in a
much more wide and sweeping form in the process of
approaching scientific problems.

Polanyi provides a paradigm case in the Stratton
experiment using inverted spectacles. The Stratton
experiment consisted of having subjects wearing spectacles

---

[26]Polanyi, _Tri-Quarterly_, p. 112.

[27]It is helpful to remember that the evaluation of
data and specific, mathematical as well as procedural
mechanisms are intimately connected with one's heuristic
vision. One's heuristic vision encompasses many specific
articulate structures but their use, significance, and
interpretation are grounded in one's intellectual dwelling place.

that inverted the image entering their eyes. After en-
countering several trying days the subjects began to
manage manual tasks with skill. Polanyi asserts that
the subjects began to reintegrate visual clues with other
sensory clues to arrive at coherent visual images of the
objects before them.

Though the image continued to enter invertedly into
the subject's eyes' the subsidiary significance of each
impression was altered so that the subjects could jointly
coordinate the impressions with those of the other
senses. The process of transforming the subsidiary indi-
cativeness of visual clues was guided by the subjects'
general vision of the nature of reality; they knew what
they should expect; the transition of the meaning of the
subsidiaries of the visual stimuli was not accomplished
formally. Polanyi generally concludes that "Every interpretation of nature, whether scientific, non-scientific
or anti-scientific, is based on some intuitive conception
of the general nature of things."[28]

Polanyi illustrates a scientific parallel to the
integrative capacities evidenced in learning to handle
problematic visual clues with Einstein's discovery of
relativity. Einstein himself recounts his own fascina-
tion, at the age of 16, with the question of whether or

---

[28]Polanyi, Science, Faith and Society, p. 10.

not a light source, if moving with sufficient speed in the same direction as the light it emits, could overtake the light itself. Einstein's intuitive heuristic vision led him to believe the answer was negative. Polanyi states that Einstein

> ...had started from the principle that it is impossible to observe absolute motion in mechanics. When he came across the question whether this principle holds also when light is emitted, he felt that it must still hold, though he could not quite tell why he assumed this.[29]

Einstein began his discovery of relativity with the formulation of a problem based upon unaccountable assumptions. They acted subsidiarily to create an image of the world as much in discord with the Newtonian tradition in physics as an inverted visual image in a right-side up world. Polanyi states that

> only a comprehensive problem, like relativity, can require that we recognize such basic conceptions as we do in learning to see rightly through inverted spectacles. Relativity alone involves conceptual innovations as strange and paradoxical as those we make in righting an inverted image.[30]

Perceiving coherent objects requires the subsidiary use of a general conception of the world; the process of scientific discovery also begins with a heuristic vision. Such a vision is applied intuitively; its

---

[29] Polanyi, Tri-Quarterly, p. 115.

[30] Ibid.

178

integrations, anticipations, and evaluations occur at a subsidiary, and therefore, informal and unspecifiable level.

Polanyi conceives of the intuition utilized in scientific discovery similar to that which functions daily in the act of perception. The concept of intuition denotes acts of integration that occur informally rather than explicitly.[31] Intuition is, of course, fallable. Different intuitive integrations referring to the same coherence will often "be of unequal value and most of them will contain but a vague or excessively distorted form of the truth."[32] Polanyi explains further,

> The intuition I have recognized here is clearly quite different from the supreme immediate knowledge called intuition by Leibnitz or Spinoza or Husserl. It is a skill of guessing right; it is a skill guided by an innate sensibility of coherence, improved by schooling. The fact that this faculty often fails does not discredit it; a method for guessing 10% above average chance on roulette would be worth millions.[33]

It is a significant point that Polanyi maintains that the intuition can be strengthened through education. The specifics of the manner in which the intuitive capacities of a scientist can, according to Polanyi, be developed will develop as we review the mechanism of

---

[31]Grene, Knowing and Being, p. 164.

[32]Polanyi, Science, Faith and Society, p. 37.

[33]Polanyi, Tri-Quarterly, p. 117.

Polanyi's theory of discovery. However, it is appropriate to indicate at this point that the intuition of a scientist can be educated through scientific study with scientists having decidedly developed intuitions. Such scientists know not only the specifics of the theories pertinent to their research but also how to apply them vectorially. They also know how to use empirical phenomena as clues rather than isolated facts in themselves. In working with such scientists, the young scientists undergoing training begin to intuitively integrate the rigorous and creative potentialities of scientific theories and the indicative possibilities of empirical data.

We can begin to understand the context of the process in which the intuition operates by analyzing the process by which coherent entities can be sought out and elucidated. Polanyi explains that the elucidation of coherent entities proceeds by way of two complimentary activities. He states;

> One proceeds from a recognition of a whole
> towards an identification of its particulars;
> the other from the recognition of a group of
> presumed particulars towards the grasping of
> their relation in the whole.[34]

These efforts broaden and deepen one another when they work together. Consider for a moment the act of

---

[34]Grene, Knowing and Being, p. 125.

180

my writing this dissertation. The whole endeavor began
with my stating a problem based upon my general understanding of Polanyi's philosophy of science as well as
my own heuristic vision. I was well versed in Polanyi's
work long before I was able to focus upon specific questions and interrelations between various aspects of his
philosophy; I had an initial grasp of Polanyi's work
that lacked comprehensiveness and unity. The increasing complexity and integratedness of Polanyi's philosophy
of science emerged as I applied my general understanding
to specific concerns and used my knowledge of specific
concerns to deepen and clarify my general understanding.
By alternating analyses of particulars, selected in
accordance with my intuitive vision of Polanyi, and applying these analyses subsidiarily to the integration of
a more detailed and cohesive vision, the coherence of
Polanyi's orientation began to emerge.

Whether the coherence one seeks is in the unity of
another man's thinking or in the ontological operational
principles unifying the patterns of phenomena we perceive,
the analytic-synthetic dynamism functions to elucidate
the coherence. Polanyi explains that the structure of the
act rather than its content provides for its effectiveness. He states,

> The concerted advantage of the two processes
> arises from the fact that normally every dis-
> memberment of a whole adds more to its under-
> standing than is lost through the concurrent

weakening of its comprehensive features,
and again each new integration of the
particulars adds more to our understanding of them than it damages our understanding by somewhat effacing their identity. Thus an alteration of analysis and
integration leads progressively to an
ever deeper understanding of a comprehensive entity.[35]

We may add that the intuition works within a con-

text of alternating analytic and integrative processes.

It is, however, most important to understand that the

intuition does not operate only during the integrative

phases of the elucidation of coherences. Polanyi opens

himself to such a misunderstanding by speaking of the

movement toward discovering and specifying coherent

entities both in terms of the alternation of analysis

and integration and intuition and imagination. Further

confusion is caused by Polanyi's lack of clarity about

the definition of his notion of analysis and his failure

to explicitly state the relationship between the two

sets of alternatives.

With regard to the first difficulty, one might ex-

pect that the term "analysis" refers to specific, ex-

plicit, intellectual processes; as we shall see, however,

Polanyi's concept of analysis within this particular con-

text is not restricted to such formalism. With regard

to the second matter, Polanyi writes that "the intuition

informs the imagination which, in turn, releases the

---

[35]Grene, Knowing and Being, p. 125.

powers of the intuition."[36] In yet other passages
Polanyi contends that "all manner of discovery proceeds
by a see-saw of analysis and integration similar to that
by which our understanding of a comprehensive entity
is progressively deepened."[37] The specific questions to
be asked are whether or not integration is equivalent to
intuition and whether or not the analysis is equivalent
to imagination. If not, how do they differ and interact? We must find our own way here.

The analytical phase of the progressive elucidation
of coherent entities refers simply to the reduction of
an initial intuitive anticipation of coherence in the
form of a problem; it is the breaking down of the problem into smaller elements that can be managed more easily.
The significant point to remember is that the original
problem is vague and does not submit to a clearly defined,
systematic reduction. The breaking down of the initial
problem must be guided by one's heuristic vision itself.
The same vision that gave rise to the problem must oversee its segmentation. Thus, the intuitive conception of
the indeterminate whole is active in what might have been
thought to be a purely deductive process.

This is not to say that formal and explicit

---

[36]Polanyi, _Tri-Quarterly_, p. 121.

[37]Grene, _Knowing and Being_, pp. 129-130.

processes do not enter into the analysis of what the scientist believes to be a coherent entity. Mathematical and logical procedures facilitate the scientist's attempt to understand the phenomena before him in a systematic way; however, the phenomena he focuses upon and the explicit procedures he accredits are contingent upon his intuitional vision of what he seeks. Polanyi maintains that "a formal system of symbols and operations can be said to function as a deductive system only by virtue of unformalized supplements, to which the operator of the system acceded."[38] Polanyi further explains, "The legitimate purpose of formalization lies in the reduction of the tacit co-efficient to more limited and obvious informal operations, but it is nonsensical to aim at the total elimination of our personal participation."[39] In essence, the formalisms of deduction in the process of scientific discovery are the concretization of informal patterns of inference; it is on the basis of this tacit coefficient that the foramlisms are given their status. The formalization of tacit processes is not merely a repacking of fixed conceptions; it is the transformation of informal and vague intimations into articulate, specific and public structures. New

---

[38]Polanyi, _Personal Knowledge_, p. 258.

[39]Ibid., p. 259.

relations between hitherto unrefined anticipations begin to emerge. Once scientists arrive at fixed conceptions they can manage, manipulate, and otherwise symbol-
ically represent large fields of data. This transformation, in its most general form, allows for the establishment and evolution of science itself.

The analytic aspect of the process of scientific
discovery operates under the guidance of the intuition
and is implemented by the imagination. The use of subsidiaries to achieve an intuitively anticipated focus
is an act of imagination. The imagination thrusts subsidiaries forward toward filling the gap between the
general use of subsidiaries in the integration of a
heuristic vision and the specific use of subsidiaries to
fulfill, in detail, the integration of the projected
coherence. The difference between the intuition and the
imagination in this case is similar to the difference
between a general knowledge of a region, its features
and its boundaries, and the exploration of the area itself. A scientist's heuristic vision must be particular-
ized and implemented if a discovery is to be achieved.
Polanyi states that,

> The questing imagination vaguely anticipat-
> ing experiences not yet grounded in sub-
> sidiary particulars [anticipations provided
> by the intuition] evokes these subsidiaries
> and thus implements the experience the

imagination has sought to achieve.[40]

The analytic process under consideration is accomplished by the subsidiaries evoked by the imagination. Numerous phenomena, elements of previous theories and particular formal mechanisms are drawn from the massive subsidiary grounds of the scientist's intellectual framework; the scientist selects and applies the subsidiaries he chooses to bring various particulars of the coherence sought into clarity. It helps us to understand the significance of the role of the imagination in focusing upon particulars of a coherence if we recognize that, for Polanyi, the only difference between an initial problem and its eventual solution is the specificity of the subsidiaries and their integrations;[41] the coherence discovered integrates much larger numbers of subsidiaries in more complex integrations than does the original problematic anticipation of the coherence.

There is no formal process by which the imagination sets about the analysis of an indeterminant coherent entity. Any formal rule reflects an inferential channel the imagination and intuition have jointly developed and accredited. The vagueness or intellectual unsettledness we may feel in considering the activity of the imagination

[40]Grene, Knowledge and Being, p. 200.

[41]Ibid., p. 202.

is a function of the asymstematic nature of the process itself. "The striving imagination has the power to implement its aim by the subsidiary practice of ingenious rules of which the subject remains focally ignorant."[42]

In sum, the intuitively guided, analytic segment of the analytic-integrative cycle sets the stage for the imagination to select subsidiaries (formal and/or informal) to break down the general scientific problem. The imagination provides not only for the particularization of various aspects of an anticipated coherence but for the subsidiary use of particulars.

On the integrative side of the analytic-integrative cycle, the imagination discerns specific parts of a coherence in an intuitively-directed context and creatively expands these elements into possible cohesive integrations. This imaginative transformation of the subsidiary character of a segment of a coherence that is in the process of being elucidated is clearly illustrated in Einstein's account of his discovery of relativity theory. Consider again the following quotation by Einstein: he states that relativity theory was discovered

> after ten years' reflection...from the paradox upon which I had already hit at the age of sixteen: If I pursue a beam of light with the velocity C (velocity of light in a vacuum), I should observe such a beam of light as a spatially oscillatory

---

[42]Grene, Knowledge and Being, p. 200.

electromagnetic field at rest....From the
very beginning it seemed to me intuitively
clear that, judged from the standpoint of
such an observer, everything would have to
happen according to the same laws as for
an observer who, relative to the earth, was
at rest.[43]

Einstein here states that he was guided by what

Polanyi would call an intuitive heuristic vision which

led him to anticipate, generally, a coherence. His in-

creasing knowledge of physics provided him with increas-

ingly complex subsidiaries that he could apply to the

refinement of his problem; he imaginatively applied his

knowledge of contemporary theories of physics in an inex-

plicit manner, a manner which attested to his own in-

articulate intelligence rather than the formal structure of

the theories themselves. Further, his last sentence

portrays his imagination at work in trying to integrate

a vision of the universe from a unique perspective where

the vectorial quality of empirical phenomena was trans-

formed; the data was seen to indicate new patterns of in-

tegration. The recognition of these new gestalten

depended upon Einstein's imaginative dwelling within the

phenomena from a perspective empirically unavailable to

him.

The thrusting forward of the imagination in search

of new possible integrations indicative of ontological

---

[43]Polanyi, _Personal Knowledge_, p. 10.

coherence is evidenced in Kepler's discovery of the eliptical orbit of Mars. His vision of a vague mechanical relation between Mars and the sun[44] made him "exclude all epicycles and send his imagination in search of a single formula, covering the whole planetary path both in its speed and shape."[45]

The efforts of the imagination are tempered by the surmises of the intuition that assess the imaginative integrations in terms of the overarching vision sustaining inquiry. Polanyi writes, "The imagination must attach itself to clues of feasibility supplied to it by the very intuition that it is stimulating, sallies of the imagination that have no such guidance are idle fancies."[46] The successfulness of the complex integrations offered by the imagination is assessed by the intuition as it attempts to judge the gains of the imagination by the general anticipations having given rise to the inquiry.

The intuitive assessment of the potentialities offered by the imagination may include the use of specific formulae. It is important to note, however, that such formulae are merely fixed channels through which inarticulate intelligence assesses phenomena. Polanyi

---

[44]Polanyi, _Tri-Quarterly_, p. 119.

[45]Ibid.

[46]Ibid., pp. 119-120.

maintains that "the mere manipulation of symbols does not itself supply any new information, but is effective only because it assists the inarticulate mental powers exercised by reading off their result."[47]

The findings of the intuition, whether they direct the scientist to reconsider his previous concerns or open vast new areas for the imagination to once again surge into the unknown, are informally grounded. Formal mechanisms may be employed to verify the final integrations. (Verification, for Polanyi, does not mean that something is proven true but rather that there is good reason to believe that an ontological coherence has, in fact, been at least partially identified.) The standards, however, that underly the evaluation of a given theory are based upon the scientist's intuitive assessment of the validity of the formal measures employed.

In the course of a discovery the intuition and imagination continuously alternate with and stimulate one another. The process of discovery is marked by many analytic and integrative efforts both of which contain intuitive and imaginative elements. We can compare the process of discovery to the creation of an arch from the known to the unknown. This comparison was suggested by G. Polya who drew attention to the paradox that although each stone

---

[47]Polanyi, Personal Knowledge, p. 83.

in an arch depends upon the others for its stability, each stone is put in separately. Polanyi responds to the paradox by explaining "that each successive step of the incomplete solution is upheld by the heuristic anticipation which originally evoked its invention: by the feeling that its emergence has narrowed further the logical gap of the problem."[48]

In summary, an analytic phase in the analytic-integrative dynamism includes both intuitive and imaginative processes. The breaking down of a problem, for example, into a number of smaller units requires not only the ability to intuitively surmise the problem itself but also the capacity to imaginatively evoke the specific subsidiaries necessary to complete such a task. The intuition can surmise relations among sets of data but it cannot create the structures necessary to reduce them. Once again, there is a difference between drawing a map and knowing how to use it. Conversely, the integration phase of the analytic-integrative dynamism employs the imagination and subsequently the intuition. In this case, the possible relations among the subsidiaries indicating an ontological coherence are effused by the imagination; possibilities are spontaneously generated by the imagination based upon the preparations undertaken in the

---

[48]Polanyi, _Personal Knowledge_, p. 128.

analytic phase. The integrative value of the possibilities is then surmized by the intuition which in its turn
spurs the imagination to continue or redirect its efforts.

We find such contentions confirmed in Polanyi's
concepts of induction and deduction. Polanyi claims that
scientific discovery has both inductive and deductive
phases but not in a classical sense. Inductive processes
are utilized by scientists but not in the absence of larger
anticipatory structures; empirical phenomena function as
clues. In integrating these clues the scientist necessarily depends upon his heuristic vision (as well as formal
processes to make information intellectually manageable).
Polanyi writes, "Empirical induction, strictly applied, can
yield no knowledge at all...."[49] "All attempts to formulate strict rules for deriving general laws from individual
experiences have failed."[50] The heuristic vision of the
scientist is necessary for Polanyi because he maintains
that science, ultimately, seeks ontological knowledge.
Consequently, observations provide only partial glimpses
of the reality being explored. Inferences derived (however one attempts to do so) from data alone are necessarily limited and incomplete.

We also find a deductive element in Polanyi's theory

---

[49]Grene, Knowing and Being, p. 41.

[50]Ibid., p. 166.

of scientific discovery. "An important degree of all discovery is deductive. For no inquiry can succeed unless it starts from a true, or at least partly true, conception of the nature of things."[51] The "conception" that Polanyi mentions is the heuristic vision. The deductive process, however, is not formally or logically controlled even though formal and logical processes may be used once the vision has, to a degree, been made articulate.

In addition, the deductive process includes the use of empirical data to specify and evaluate expectations and to provide problematic areas for further consideration.

There still remains another aspect to Polanyi's concept of scientific discovery; it refers to the phases that occur in the overall process of discovery. Polanyi occasionally refers to the work of Poincare who indicated four phases of stages of discovery.

Polanyi adapts the four stages--Preparation, Incubation, Illumination, and Verification--to his theory of scientific discovery, but he also refers to scientific discovery in terms of the imagination and intuition.[52] These aspects of discovery are neither explicitly interrelated by Polanyi nor are they described in terms of the fundamental tacit triadic structure of knowledge. The following

---

[51]Grene, _Knowing and Being._ p. 130.

[52]Polanyi, _Personal Knowledge_, p. 195.

section attempts to unify the above considerations in a cohesive overview of the process of scientific discovery as implied by Polanyi.

We can find evidence of the four stages of discovery in the following quotation by Polanyi:

> A problem fit for inquiry comes to the scientist in response to his roaming vision of yet undiscovered possibilities. Having chosen a problem, he thrusts his imagination forward in search of clues and the material he digs up--whether by speculation or experiment--is integrated by intuition into new surmises, and so the inquiry goes on to the end.[53]

The Preparation stage of discovery refers to the scientist's attending to the phenomena available to him in order to distinguish a problematic situation as yet unseen. The nature of the problem can be in almost any form, from the observation of anomalies to the disjunction of two theories. During this stage the researcher (A) uses his own vision of the world (inclusive of specific theories, rules of logic, empirical data, etc.) (B) to form a growing recognition and appreciation of a problem (C). This stage is predominantly intuitional as it must integrate numerous factors, some of which are themselves indeterminate, i.e. the scientist's personal world view and functional understanding of formal theories. The imagination is active in that it can thrust subsidiaries into the contemplative

---

[53]Grene, <u>Knowing and Being</u>, p. 202.

context of the intuition. It is the intuition, however, that must ultimately envision a worthy problem. The scientist's general intuitive assessment of his own capacities, the resources available and the time required also enter into his degree of involvement with a given problem.[54,55]

The second stage, the Incubation of the problem, is one in which the researcher first begins to attend directly to the specifics of the problem itself. Pre-paration discovers the problem while Incubation inform-ally uses the problematic empirical or theoretic data, subsidiarily, to discover the resolving coherent entity. The tacit structure of this stage consists of the scien-tist (A) subsidiarily dwelling within the vision of the problem with its informal inclusion of current scientif-ic theories and vectorially charged empirical data (B), focusing upon the hidden coherence (C).

This phase of scientific discovery is predominantly inarticulate; it is a spontaneous activity over which

_____

[54]The discovering scientist must begin his work not only with an intuitive surmise of the ontological principles active in given phenomenal fields but also a "strategic intuition" of the resources, time and personal intellectual capacities needed to pursue the problem to its conclusion. An underestimating "strategic intuition" might well assure mediocre results from outstanding intellectual gifts and an overestimating integration of the factors noted might well assure wasted or confused efforts.

[55]Grene, Knowing and Being, p. 202.

the scientist has little conscious control. Polanyi
maintains, "The fact that our intellectual strivings
make effective progress during a period of Incubation
without any effort on our part is in line with the
latent character of all knowledge."[56] This latent
character refers to the informal efforts by which an
individual attempts to span the gap between his intuitive vision of a coherence and his ability to make subsidiaries function to bring that focus into clarity.

The question may arise as to why the Incubation
of a problem is largely unconscious. A possible explanation may lie in the fact that the Incubation stage is
predominately imaginative; it is a questing for particulars that can add more specificity to the problem and a
thrusting forth of possibilities for further intuitive
assessment. It is only when the scientist begins to
integrate the fruits of the imagination that the work
of the imagination becomes an object of contemplation.
It is largely unconscious because it consists of the
first probes of the mind that have yet to be confined
systematically to the parameters and rules of language.
Polanyi explains (although without reference to the
stages of discovery here being considered) that

> Scientific discovery which leads from one...
> framework to its successor, bursts the bounds

---

[56]Polanyi, Tri-Quarterly, p. 100.

> of disciplined thought in intense if not
> transient moment of heuristic vision [The
> close of stage of Preparation]. And while
> it is thus breaking out, the mind is for
> the moment directly experiencing its content
> rather than controlling it by use of pre-
> established modes of interpretation [The
> height of Incubation].[57]

The phrase "the mind is for the moment directly
experiencing its content" refers to the contention that
at this phase of discovery, the restrictions of systematic procedures, even those of language, are transcended;
the researcher transcends the symbolization of language and relies on the presuppositional intellectual
ground that underlies his acceptance of even the most
formal and explicit rules and standards of science.

The process of imaginatively working with subsidiaries at this stage is most dynamic as the subsidiaries drop the static character they have in explicit
or routinely used systems of interpretation. The subsidiaries of past experience, empirical data, the presuppositions of the theories which the scientist had
accepted at the outset weave together plastically, with
fluidity. Einstein's description of his imaginative
journey on the edge of a beam of light illustrates such
a phase. Polanyi writes that "such activity dissolves
the screen [of symbolic representations], stops our

---

[57]Polanyi, _Personal Knowledge_, p. 196.

movement through experience and pours us straight into experience; we cease to handle things and become immersed in them."[58]

It is at this point that a well chosen problem will allow the intuitional anticipations of the researcher a communion with the reality he seeks.[59] It is at this point where the researcher can find the imaginative implementation of his heuristic vision and discover contexts for the data he contemplates, contexts that indicate a rational coherence.

This process does not necessarily involve a single unitary effort; the incubation period is constantly redirected by intermediary intuitional surmises. These surmises integrate the thrusts of the imagination in accordance with the anticipations of the heuristic vision.

As the imagination comes closer to leaping the gap between the researcher's antecedent knowledge and

---

[58]Polanyi, _Personal Knowledge_, p. 197.

[59]The reality here indicated refers to the orderliness and lawfulness that the researcher attempts to eventually represent in a rational form. His communion with reality refers to the degree to which his anticipatory intuitions and imaginative use of subsidiaries accurately parallel the actual ontological coherence and its operational principles as it creates the phenomena having given rise to the problem under study.

an increasingly comprehensive intuition of the coherence sought, the researcher's own activity has given
him greater facility to rapidly accomplish the final integrations. This stage of discovery often results in a
more concentrated effort in a "self-accelerating" manner
so that the final discovery may come in a seemingly
sudden flash.[60] This is the stage of Illumination.

In the Illumination stage, a scientist (A) brings
his imaginative output and anticipatory intuition (B) to
bear upon the explication of the principles at work in
the ontological coherence he has studied (C). Polanyi
writes that "A final intuition is a claim to discovery
and is thus at the point of being proved true or mistaken,
while a problem--a feat of anticipatory intuition--can
usually be justified only after a long and uncertain
pursuit."[61]

The stage of Illumination generally integrates the
subsidiaries it handles into articulate structures. The
Incubation stage was said to "burst the bounds of disciplined thought" and now we may add that the Illumination
stage reintroduces the incubational findings into systematic and disciplined form.

---

[60] Polanyi, The British Journal for the Philosophy
of Science, p. 100.

[61] Grene, Knowing and Being, p. 202.

The question arises as to how formal, explicit, and logical procedures and standards play in this entire process. The answer is quite simply that the formalisms usually associated with the rigors of scientific inquiry are employed increasingly during the last part of the Incubation stage and through the Illumination stage as called for by the intuition. Their role is ancillary; they are complex tools applied by the scientist as he envisions a goal and the utility of a given procedure. Explicit, mathematical procedures are extremely helpful to the scientist because they provide fixed channels for integrating large amounts of data and permit the clarification of numerous relationships and aspects of the field of phenomena that were perhaps never considered. Polanyi concludes that,

> both the first active steps undertaken to solve a problem and the final garnering of the solution rely effectively on computations and other symbolic operations.... However, the intuitive powers of the investigator are always dominant and decisive.[62]

Formal structures are employed in much the same way everyday language is employed. The inarticulate powers of intelligence imbue every utterance, every graph and statistic with a relative degree of meaning and significance. The intuition provides the context

---

[62] Polanyi, _Personal Knowledge_, p. 130.

for selecting procedures just as it provides the context
for a choice of words; the imagination presents the intuition with ways of meeting contextual demands just
as the imagination searches for (or creates) the word(s)
needed to explicate the findings of the intuition (see
Chapter III).

The last stage of scientific discovery is known as
that of "Verification." In the process of verification
the researcher uses his subsidiary knowledge of his own
general vision of reality, his discovery, as well as his
general knowledge of the procedures for the reduction,
if not elimination, of extraneous variables, to create
an experimental procedure. The researcher should expect
that the empirical data he derives from experimentation
will correspond to his predictions based upon his discovery. We may generally say that, in the context of
the tacit knowing triad, the researcher (A), with universal intent, applies his general conceptions of science, his
scientific discovery and knowledge of particular scientific and mathematical procedures (B) to focus upon the cor-
relation between predictions and experimental results (C).

The confluence of expectations and empirical results, however, does not in any way prove that a scientific discovery in fact discovers an ontologically coherent orderliness. Further, the divergence of expectations and observations does not in any way disprove

or definitively refute a claim of discovery. In the former case, future expectations may not find similar confirmation, and in the latter case, future research may reveal that previous anomalies were purely circumstantial, i.e. the result of inadequate research design.[62] Both verification and falsification require that scientists make judgments about the relative significance of any experimental attempt at verification or falsification and to that extent, "verification and falsification are both formally indeterminate procedures."[63]

The general assumption underlying the idea that verification can proceed in a purely formal manner is that one assesses a theory through reference to clear and explicit facts. We have already seen, however, that "Things are not labelled 'evidence' in nature, but are evidence only to the extent to which they are accepted as such by us as observers."[64] Continuing scientific controversies revolve around the problematic "factuality" of life after life experiences, extrasensory perceptive capacities, accupuncture treatment and homeopathic therapy.

---

[62]As noted previously, D.C. Miller recorded anomalies to Einstein's relativity theory for 25 years only to have the anomalies the possible result of "statistical flctuations and temperature effects" (Personal Knowledge, p. 13). For a further discussion, see Chapter I.

[63]Polanyi, Tri-Quarterly, p. 111.

[64]Polanyi, Personal Knowledge, p. 30.

The embattled sides draw different conclusions about these issues because their heuristic orientations would have them dispute whether facts are indeed facts or fictions. Polanyi maintains that, "Of this responsibility we cannot divest ourselves by setting up objective criteria or verifiability--or falsifiability, or testability, or what you will."[66] In the end, even our rules for verification fall back upon standards that were accredited as such in the absence of the rules and standards themselves. Falsification also requires that we begin with the assumption of certain fiduciarily-rooted conceptions whether we call them empirical observations or test statements.

In so far as a scientist attempts to introduce a new rational coherence having ontological status to the scientific community, he appeals to the basic and fundamental standards of his peers rather than their own specific conceptions of an issue. The discovering scientist, disciplining himself by standards to which he would hold any other scientist accountable and accordingly revealing what he believes to be a new aspect of reality that demands him to accept its existence, asks the members of the scientific community to review his findings with the same rigor and universal intent to which he

---

[66]Polanyi, _Personal Knowledge_, p. 64.

commits himself. He appeals to the common grounds of scientific tradition as the basis for the acceptance of his discovery. "On the grounds of the self-command which bound him to the quest of reality he must claim that his results are universally valid: such is the universal intent of science." [67]

---

[67]Polanyi, <u>Tri-Quarterly</u>, p. 122.

Chapter V

SUMMARY AND IMPLICATIONS

Summary

Polanyi maintains that scientific theory should
not refer to only that which can be clearly observed
or that which has been couched in logical terms.
Scientific inquiry itself, he believes, is based upon
a search for what he calls the "rationality of
nature": those lawful ontological principles that
govern that which we perceive as well as that which we
have yet to observe. Knowledge of such principles
thus yields anticipations of further contact with the
reality underlying new and diverse fields of inquiry.

Scientific knowledge is therefore indeterminate
both in terms of its object (the principles that
extend beyond the limitations of present data) and its
heuristic anticipations. Polanyi's ontological
referent for science necessarily places science upon
a fiduciary foundation. Neither formal mechanisms
nor logical operations can, themselves, formulate
scientific theories or assess the heuristic power of
a given theory. Thus, the progress of science depends

upon the a-critical acceptance of certain primary premises about the nature of science and object of science. Such preconceptions are partially shared by the scientific community and are partially the personal interpretations of individual scientists.

It is important to note that the premises of science are largely a-critical as they are functional interpretative systems rather than static objects. Their meaning in the conduct of inquiry derives from their vectorial function rather than their isolated content. They are the background against which factuality appears; they cannot be compared with "the facts" as they operate to establish the facts.

Consequently, Polanyi's concept of objectivity rests neither solely upon logical nor solely upon empirical foundations. He maintains that a theory may be said to be objective to the degree that it mirrors the governing principles at work in the field of study. Thus, the objective value of a theory cannot be wholly formally assessed even though numerous mathematic and technical procedures can be employed; there is simply no way to formally determine the relation between a theory and an ontological

principle which is phenomenally indeterminant. Polanyi contends that the value of a given theory can ultimately be determined only by the future confirmation of its anticipations. Thus, beyond all the formal evaluative mechanisms that can be used, a scientific theory must be judged, in part, by the presuppositional visions scientists have and their willingness to pursue their conclusions.

Such assessments are held in check by two mechanisms, one social, one personal. The first is conviviality; it refers to the fact that the individual scientist lives within a community of scientists whose opinions affect publications, research grants, professional positions, etc... The second is "universal" intent; it refers to the scientist's responsibility to pursue his research and confer his judgments as his vision of reality would have him hold as universally necessary. In this sense, scientific theory can be personal (in its grounding in the personal commitments of the scientist) but not subjective (in its rejection of idiosyncratic desires or preferences). As we noted in Chapter I, "In so far as the personal submits to requirements acknowledged by itself, it is not subjective; but in so far as it is an action guided by personal passions [personal commitments to and beliefs about the nature of reality and the specific area of research], it is not objective either. It transcends the distinction between subjective and objective."[1]

Polanyi explains that the ontological referent of science serves as an external pole, supporting the growth of scientific knowledge. He states that we trust our presuppositions and theories,because "We believe that their manifest rationality is due to their being in contact with domains of reality, of which we have grasped some aspect."[2]

The "grasping" of some aspect of reality consists in the discernment of gestalten that are indicative of ontological principles.

The question as to how such gestalten, or patterns of phenomena, can be distinguished from purely coincidental patterns cannot be accomplished formally. No mathematical computation itself can discern which, from a variety of possible correlations, indicate ontological principles. Polanyi states that

> Mathematical reasoning about experience must include, besides the antecedent non-mathematical finding and shaping of experience, the equally non-mathematical relating of mathematics to such experience and the eventual, also non-mathematical, understanding of experience elucidated by mathematical theory.[3]

---

[1]Michael Polanyi, _Personal Knowledge: Toward a Post-Critical Philosophy_ (New York: Harper & Row, 1964), p. 104.

[2]Ibid.

[3]Marjorie Grene, _Knowing and Being_ (Berkeley: University of California Press, 1974), p. 179.

Hence, the discovery of ontological principles, although employing highly developed mathematical processes, relies fundamentally upon the informal insights of the scientist, insights developed and sustained by his heuristic presuppositional framework, his heuristic vision.

This informal and personal shaping of experience undergirding and directing the use of mathematical formulae constitutes an aspect of knowledge which must be personally sustained by the scientist. To this extent, knowledge does not indicate that one assimilates a given content, but more, that one has developed the capacity to use such concepts, theories, ideas, etc....to actively interpret new and diverse experiences. Thus, to have knowledge of a scientific theory implies that one has the personal heuristic vision and can employ the theory as a probe to integrate the diverse elements one encounters in one's field of research.

Further, each scientific theory embodies, in its form and structure, premises that were instrumental in its development. To understand the theory is in part to assume the presuppositions that underly it, if only to reject them later as inadequate according to one's own vision. This implicit as opposed to explicit foundation of scientific theory allows for the application (and, perhaps, eventual transformation) of a given theory to varied areas of research.

To the extent that all scientific theory is embedded in an unspecifiable substructure, all scientific knowledge may be said to have a logically unrepresentable tacit element. This cognitive underpinning is dependent upon a scientist's personal assimilation of the presuppositional vision that vectorially charges data and that accredits mathematical relations as significant.

Such tacit knowledge of a scientific theory shapes anticipations and identifies the possibility of fertile problems for research not necessarily directly deducible from the theory itself. The history of science clearly demonstrates that even though great discoveries may have depended upon previous research and theory, the discoveries themselves could not have been formally derived from such antecedent factors.

Polanyi maintains a cognitive element in the individual that can make use of unspecifiable data, the element that permits the development and application of unspecifiable presuppositions, is evidenced even in the act of perception. He notes Gestalt psychology experiments in which subjects were found to perceive certain symbols only in terms of their effect upon the subjects rather than their specific perceptual identity. The symbols were functional and their identity was indistinct; they were subsidiary to the focus of the subject's attention. The effects of the symbols were specific and identifiable; they were focal. Polanyi maintains that this simple experiment illustrates an aspect, a fundamental quality of knowledge: all knowledge depends upon the subsidiary awareness of some entities to focus upon others. All knowledge is tacit or tacitly-rooted.

In the case of scientific knowledge, presuppositional commitments occupy the subsidiary position while the specific content of theories may occupy the focal position.[4] The latter devoid of the former are merely marks upon a sheet of paper signifying nothing. In so far as a theory is said to have meaning or significance, it is believed to be indicative of a reality beyond itself. In so far as one applies a given theory to new and varied areas of research, one has grasped its general presuppositional base, its interpretive tenets. The subsidiary use of the premises of a theory is itself, non-deductive as the presuppositions'

---

[4]It should also be noted that when a scientist applies a theory to a given set of data to discern gestalten indicating ontological coherences, the theory occupies a subsidiary position. For a more detailed discussion see Chapter IV.

vectorial qualities are quite distinct from their specific content.

In general, Polanyi asserts that all knowledge has a triadic structure. The "tacit knowing triad," as he calls it, is composed of a person (A) who used subsidiaries (B) to focus upon an object (C). The question arises as to the nature of the person in such a framework of "personal knowledge." Polanyi's concept of indwelling provides insight into Polanyi's response to this inquiry. He notes that as one applies subsidiaries, they tend to provide a context, an intellectual dwelling place, from within which one can attend to particular objects and/or events. He further remarks that one's body almost always occupies a subsidiary position which one thus comes to know as one's own.

The subsidiary position of the body provides, as it were, an existential dwelling place. Polanyi, however, does not exclude the "interiorization" of yet other objects as subsidiaries from which an individual can attend to focal entities. All objects occupying the B position become part of a complex subsidiary network. Polanyi concludes that consciousness is bi-polar and intentional in that interaction with the world requires both subsidiary and focal elements where the former bring the latter into focus.

It might appear that Polanyi attempts to use our indwelling experience of our bodies as a means of overcoming the classical Cartesian mind/body dualism. More specifically, one could conclude that the mind and body merge in the subsidiary position where one existentially experiences one's body as a point of insertion into the world.

However, Polanyi maintains that knowledge has a triadic structure which not only requires subsidiaries and a focus, but also a person who brings the former to bear upon the latter. Consequently, it must be asked where subsidiaries end and the mind begins. If our bodies are said to occupy a subsidiary position, must we conclude that neural activity and electro-chemical cortical processes are also instances of subsidiaries rather than processes of a mind? Polanyi answers that indeed we must. He contends that cortical processes cannot be understood properly unless we grasp the idea that the experience of a neural or cortical process, its potency, meaning and context, are distinct from a detached analysis of the process itself.

> ...neural functions supply...signs but
> they do not supply their interpretation.
> Since this interpretation forms no part
> of the nervous system, the system
> cannot be said to feel, learn, reason,
> etcetera. These are experiences or
> actions of the subject using his own

neural processes.[5]

Thus, Polanyi concludes that the mind and the body are not two aspects of the same thing but are, in fact, separate entities. The ontological status of the Polanyi concept of the mind can be traced further through his account of its interaction with the body.

It is clear that Polanyi believes that the mind subsidiarily applies the body (including neural and cortical mechanisms) to focal objects. Polanyi asserts this interaction is an instance of what he calls the principle of "dual control." The principle of dual control asserts that natural laws operating at one level, i.e., physical and chemical laws, can create a complex field condition that will allow for other principles to emerge, i.e., biotic laws. Once the latter have emerged they obey the former, but they also operate according to their own mechanisms. Polanyi explains that biological organisms depend upon physico-chemical laws but also embody aspects that cannot be ascribed to such laws, i.e., sentience, consciousness, and in man, reason. Polanyi writes

> Mental principles and the principles
> of physiology form a pair of jointly
> operating principles. The mind relies
> for its workings on the continued
> operation of physiological principles

---

[5] Grene, Knowing and Being, p. 202.

but it controls the boundary conditions
left undeterminedly by physiology.[6]

Consequently, Polanyi maintains that all thought
is incarnate, but that it is not physio-chemically
determined. Polanyi attempts to explain the presence
of contemplative principles such as the rule of logic,
the restrictions of rational thought and the commitment
to universal intent, as a result of complex physio-
chemical and biotic field conditions; he attempts to
introduce the operational principles at work in the
mind as a subclass of physio-chemical processes. Thus,
when we ask where new levels of principles come from,
Polanyi answers that they are physio-chemical in
origin. His concept of a separate mind dissolves as
the mind itself is a physio-chemical, biological structure wholly determined by physio-chemical laws. The
presence of higher order interpretative principles is
not accounted for. Clearly, Polanyi's account of the
mind/body relation is problematic. He attempts to explain the rise of the human mind by referring to operational principles that transcend physio-chemical laws.
However, these transcendent operational principles are
themselves materially circumscribed. Consequently,
Polanyi gives us no indication why he would ascribe
such properties as sentience, consciousness or the

---

[6]Michael Polanyi, "Logic and Psychology," American
Psychologist, 23 1 (January, 1968): 40.

aforementioned characteristics of the human mind, to certain physio-chemical configurations, while denying them to others.

Despite this serious dilemma, Polanyi does offer a consistent epistemological model when we simply accept the presence of such contemplative structuring principles as mentioned. He asserts that the towering intellectual superiority of man over the animals is due to his capacity to develop formal, _articulate_ structures. This is in keeping with his emphasis upon the tacit aspect of the knowing process as all articulate structures (all language, mathematic and otherwise) are subsidiarily-rooted. To generate and interpret language necessitates that one grasp its underlying content, aims and grammar.

Polanyi distinguishes three types of tacit intellectual capacities that are evident in animals, but which are so highly developed in man as to provide a groundwork for language.

Polanyi believes that human intelligence is con-tinuous with that of animals, but he does not provide a suitable framework for the origins of the demands of logic, reason and universal intent since he contends that their functioning is _not_ physio-chemically determined.

These capacities enable animals and human beings

(to a greater degree) to acquire the subsidiary use of entities in a variety of contexts. Polanyi explains that 1) "trick" learning enables a subject to recognize a means-end relationship between two entities, 2) "sign" learning enables a subject to use one entity to foretell the behavior of another entity, and 3) "latent" learning enables a subject to heuristically apply past experiences of a given entity to unprecedented situations. In all three cases the learner must subsidiarily incorporate one item in order to attend another.

This innovative step to the subsidiary usage of an object is inarticulate in animals and can be pre-articulate in man; it is neither public nor explicit even though once the innovation is made it can sometimes be logically integrated with antecedent knowledge. Thus, the learner must ultimately learn the subsidiary use of an item for himself. This situation is similar to the construction of an arch which requires a scaffolding only until the arch is complete. Upon the completion of the arch, the scaffolding (in this case, the first tentative leaps across the logical gap from the known to the unknown) is cast aside so the arch can stand alone. (In this case, integrated knowledge.) The tacit foundations of articulate knowledge are intellectually invisible. When explicit knowledge is applied or assessed the researcher or evaluator is, similarly, tacitly active.

The acquisition of language requires a complex assimilation of numerous subsidiaries that govern the content and form of language. The use of a single word, i.e., "Man," requires that an individual

1) derive the general attributes of a category,

2) identify the characteristics of a given entity,

3) judge whether or not the attributes of a given entity constitute an instance of the category in question, and

4) appraise his own skill in performing such functions. These skills are non-specifiable and any standards that may be publically applied rely upon a non-specifiable assessment from which the standards are derived.

Polanyi identifies the components of such informal processes as the imagination and the intuition. It is the imagination that searches through our memories for a forgotten word; it is the imagination which mobilizes the particulars of a situation into useful subsidiaries for finding the word. It is the intuition that sets the imagination to its task; it is the intuition that provides the context requiring the word and confers judgment on its acceptability. This is but a minor example, but this general structure could be enlarged to account for the unfolding of grammars, explicit systems of thought and scientific discoveries. Generally, the intuition integrates subsidiaries into a heuristic framework while the imagination weaves particular subsidiaries in accordance with and, as judged by, the intuition. Polanyi states,

> It is the intuition that senses the presence of hidden resources for solving a problem and which launches the imagination in its pursuit. And it is the intuition that forms there our surmises and which eventually selects from the material mobilized by the imagination the relevant pieces of evidence and integrates them into the solution of the problem.[7]

The generation of explicit structures enables the pre-articulate powers of the intuition and imagination to greatly expand their scope and complexity. The explication of the workings of the intuition and imagination into fixed, publically available symbols, permits the isolation, exploration and formal analysis of such workings. Consequently, the implications of an individual's conception can be made accessible to others and new guiding insights can be provided for others' imaginations.

We may understand the whole process of scientitic discovery as an activity guided and enacted by the interplay of the intuition and imagination. Scientific discovery begins with a set of presuppositions that the

---

[7]Polanyi, "Logic and Psychology," p. 42.

intuition integrates into a vision of the world and uses to generate anticipations about the nature of a given field of inquiry. The intuition provides a general heuristic framework which underlies and accredits even the most formal of procedures and theories. Further, it envisions various areas of a given field that should be studied, as well as what should generally be expected. Once research has begun the scientist will not only observe phenomena, he will search for certain kinds of phenomena; he will view empirical data as clues to a hidden coherence, the barest outlines of which his heuristic vision traces.

The anticipatory power of the intuition rests upon the correspondence between one's heuristic vision and the operational principle at work in the field being studied. To this extent, a scientist may assume that his vision (which undergirds and accredits the scientific theories he employs) should generally parallel in its structure the functioning of an operational principle in its ontological context. Futher research particularizes general anticipations and may allow for the development of a scientific theory.

Numerous formal mathematical and procedural guidelines are indeed essential to research, but such items are embedded in the scientist's heuristic vision and derive their status therefrom. It must also be remembered that the application of antecedent theories to present inquiries, which may eventually result in a scientific discovery, is not usually a purely deductive process. It most often requires conceptual innovation, procedural and mathematic judgments and the creative selections of research topics. Polanyi maintains that there is a logical gap between the known and the unknown. He contends that the heuristic vision, through the intuition, can subsidiarily integrate available information into an anticipation of the far side of that logical gap. Polanyi writes that,

> true discovery is not a strictly logical performance, and accordingly, we may describe the obstacle to be overcome in solving a problem as a 'logical gap' and speak of the width of the logical gap as the measure of the ingenuity required for solving the problem.[8]

It is important to note that Polanyi does not conceive of the intuition as an immediate type of knowledge. It is an extension of an "innate sensibility" of coherence based upon our capacity to integrate subsidiaries as they are functioning in the object we are studying. The problems posed by the intuition (which, once again, incorporate numerous explicit elements) are composed of nexuses of subsidiaries that the scientist

---

[8]Polanyi, Personal Knowledge, p. 123.

has brought to bear upon a focus, a focus which is as yet only vaguely outlined by such subsidiaries. Research may reveal problematic aspects of the initial vision and may require that some premises be refined or refuted.

Polanyi explains that the object of inquiry is brought into focus by two alternating, complementary activities: analysis and integration. The former refers to the reduction of a whole into its parts and the latter refers to the integration of the parts with respect to the whole. Although Polanyi does not specify how these processes are related to the intuition and imagination it is clear that both the intuition and imagination are active in each of the complementary activities. The analysis of a given problematic situation begins with an intuitive surmise that directs the imagination to find subsidiaries (including explicit formulae and scientific theories) to discern the specifics of the object in question. The integration phase of the cycle begins when the imagination attempts to weave new and varied sets of subsidiaries which are in turn judged to be more or less suitable by the intuition. This alternation continues numerous times as the focal object assumes more detail and clarity. It is apparent that Polanyi subscribes to neither a classical inductivist nor classical deductivist concept of discovery.

Lastly, Polanyi notes the four stages of discovery identified by Poincare. However, Polanyi does not specifically indicate their relatedness to his concepts of tacit knowing, imagination and intuition. The four states are Preparation, Incubation, Illumination and Verification. They can be consistently interpreted within the contexts Polanyi provides.

In the Preparation stage the scientist brings his heuristic vision, inclusive of numerous explicit structures, to bear upon a given area of research so as to outline a problem. The intuition provides the context for the imagination to generate a possible problem through the manipulation of subsidiaries. The problem is then assessed by the intuition.

The Incubation stage is largely imaginative. It mostly consists of the researcher imaginatively weaving sets of subsidiaries to attend to the problem outlined by the intuition. Numerous cycles of integration and analysis refine and direct imaginative efforts.

The Illumination stage is predominately intuitional. In this stage the scientist brings his intuition to bear upon his imaginative thrustings so as to delineate and specify the working of a once problematic, "empty" focus. It is at this stage that the preconceptual effusions of the imagination are brought to a formal, conceptual level.

The last stage, Verification, is a balance of intuitional and imaginative functioning. Here, the researcher will, with universal intent, bring his primary conceptions of the world and science to bear upon his discovery and any empirical data available. It is possible that the discovery will appear, in the light of the scientist's personally accredited standards, false, inadequate, in need of refinement or accurate to ontological reality. It is also possible that the specifics of his discovery will have sufficient weight to demand the revision of the original presuppositional vision of reality. In any case, the assessment of the discovery is "formally indeterminant." Anomalous data may one day prove a special instance of the theory or perhaps a simple experimental error. Consistent data may be the result of the limitations of the cases studied. In the final analysis, the researcher will assess his discovery with reference to standards and measures he believes to be worthy of universal acceptance. He thus brings his work, claiming its universal validity, before the scientific community.

Implications for the Scientific and
Educational Communities

Polanyi recognized many of the implications of his theory of science. He is, perhaps, best known for his writings on the implications of his philosophy of science for the socio-political structure of the scientific community. There are numerous implications of Polanyi's general philosophy in such diverse fields as theology and biology.[9] The purpose of this section, however, is to outline some of the implications of Polanyi's epistemology for the scientific and education communities as they are the groups most directly affected. The implications for the scientific community will consist of a systematic extension of Polanyi's emphasis on the ontological referents of science while the implications for the educational community will come more directly from Polanyi's concept of the active knower in the known. It must be emphasized, however, that the ontological and cognitive aspects of Polanyi's epistemology are interdependent. Either one can be emphasized, but neither one

---

[9]For example, see "The Triadic Structure of Religious Consciousness in Polanyi" by Robert Innis, Thomist, July 1976, pp. 393-415, or the essay on biology in Toward a Unity of Knowledge by Marjorie Grene, and published by the International Universities Press, 1969.

can be separated or detached from the other.

We have seen that Polanyi argues that science does
and should seek to refer to ontological reality, for it
is the correspondence between scientific theory (with its
presuppositional substructure) and ontological reality
that permits scientists to anticipate fruitful results
from the problems they pursue. It is thus consistent
for Polanyi to claim that both scientific discovery and,
to a lesser degree, assessment are formally indeterminate procedures. This is not to say that the formalisms of different approaches to science are not valuable for they are invaluable; it is to say that such
formalisms facilitate the judgment of the scientist who
accredits the formalisms' functions and interprets their
significance in the light of his heuristic vision of the
ontological principles at work in the area being researched.

Consequently, Polanyi's epistemology implies that the
concepts of objectivity underlying various philosophies
of science including logical positivism, refutationism
and verificationism are inadequate and misleading.
They all attempt to develop specific, public procedures--
whether through wholly prescribed research methodologies
or logical transcriptions--that eliminate the judgment
of the individual. A positivistic approach to science
mistakes the clues to a systematic reality for the

referents of science. The heuristic functions of theories in the anticipation of future fruitful areas of inquiry are thereby reduced if not eliminated. Similarly, attempts at the hard and fast refutation or verification of scientific theory are based only on partial and incomplete information about the ontological principles operating in a given field of research; formal procedures may, in fact, settle the issue ninety-percent of the time, but great discoveries in science are very often the result of one scientist seeing a coherence where his peers saw none, or vice-versa. The anticipatory vision that guides scientific researchers rests upon the notion that present knowledge partially reveals ontological principles which will reveal themselves further with extended research. This heuristic vision, dedicated to the overarching laws at work in a field of study, may set aside anomalies and discount verifications.

Further, it may be argued that the significance of any particular finding must be assessed by individual scientists both for its structure and content as well as its possible utility as a basis for future research. The vast majority of scientists within a given field may well share common goals and standards, but they cannot ursurp the responsibility of each scientist to review and evaluate the significance of any given scientific theory;

the ultimate authority in science, for Polanyi, is the individual scientist who commits himself to the ideal of universal intent. To centralize the authority in science by instituting a given philosophical conception of inquiry or by standardizing any particular sets of levels of acceptability or by selecting a group of leading individuals to oversee research, would destroy the very essence of science: the individual's striving for and responsibility to the ontological principles governing the course of natural events as best as he can envision them. Each member of the scientific community evaluates the work of all the others in accordance with his vision and consequent standards (many of which he may well share with the vast majority of other members). Herein lies the basis for Polanyi's argument for the freedom and authority of the individual scientist. He summarizes,

> A community which effectively practices free discussion is therefore dedicated to the four-fold proposition (1) that there is such a thing as truth; (2) that all members love it; (3) that they feel obliged and (4) are in fact capable of pursuing it.[10]

Polanyi hopes that the scientific community will evolve into "A Society of Explorers," a society that recognizes the personal responsibility and freedom of scientists in their research. The Society of Explorers

---

[10]Michael Polanyi, <u>Science, Faith and Society</u> (London: Oxford University Press, 1946), p. 71.

would be bound by a commitment to a common tradition of scientific knowledge and understanding which they may transform as their findings would require them to.

Herein lies one of the major educational implications of Polanyi's concept of knowledge. Future members of such a society of explorers would begin their initiation through an educational process which steeped them in scientific traditions. However, it is important to note that Polanyi does not conceive of tradition as static sets of prescribed beliefs. He maintains that just as there is an ontological reality that governs the specific data we gather, so the scientific community is generally united in a conception of its goals, achievements and dispositions that overarches the specifics of any theory or set of theories. Polanyi explains,

> To understand science is to penetrate to the reality described by science; it represents an intuition of reality, from which the established practice and doctrine of science may be regarded as a much simplified repetition of the whole series of discoveries by which the existing body of science was originally established.[11]

The work of previous generations of scientists provides the context for the questions, interests and general intellectual development of future members of the society long before the scientist to be has specific

---

[11]Polanyi, <u>Science, Faith and Society</u>, p. 45.

knowledge of the meaning of science itself. For example, most of us learned that the earth travels about the sun, that a ball falls because of gravity, that time ceases at the speed of light. Many of us have pondered these lessons and very vaguely wondered about questions we might ask and ideas we might pursue. However, very few of us know the mathematics of Copernican theory lest we consider Einstein's work; we take these basic conceptions of the world on the basis of the authority of our teachers. It is only by virtue of this faith that we could be initiated into the foundations of modern science. Thus, an introduction to science requires faith in authority and tradition.

The introduction to the specifics of modern scientific knowledge begins with the novice's assimilation of the presuppositions of a modern scientific conception of the world. These premises serve as integrative guides rather than explicit rules. Polanyi asserts,

> The premises underlying a major intellectual process are never formulated and transmitted in the form of definite precepts. When children learn to think naturalistically they do not acquire any explicit knowledge of the principles of causation.[12]

The student is introduced to science as a tradition, as a perspective, as a way of viewing the world which has

---

[12]Polanyi, <u>Science, Faith and Society</u>, p. 42.

a significant a-critical foundation. The educator's task is not so much to have the student of science adopt a critical attitude but to equip him with the subsidiary context through which he can eventually address the problems posed by modern science. To this degree, the teacher, as the purveyor of tradition, exercises responsible judgment for the conceptions he presents as a model for his students. The student, in observing his teacher, employs the presuppositions of science and attempts to understand the principles operating in the process of judgment. The student attends to the teacher's instruction with the faith that initiation into the perspective the teacher represents will one day provide a vision that will be borne out by the specifics to be observed in the future. This faith that this vision will reveal new and unexpected aspects of reality underlies the student's daily efforts. The science teacher presents the student with a presuppositional image of reality, as the scientific community apperceives it. The specifics of scientific theories the student eventually considers are useful to the student only to the degree that he has the subsidiary foundation that provides for their context and relative degree of value.

Let us consider the Copernican Revolution from a pedagogical perspective. The mathematics necessary to understand Copernican and even Ptolemaic theory are beyond the sophistication of most college graduates lest we consider sixth graders. We are thus faced with the decision of either instructing students in, presumably, a heliocentric vision of our solar system or deferring the issue to such time as the student, properly mathematically equipped, can decide the issue on the basis of his own critical judgment. Delaying the presentation of heliocentric conceptions would defer the study of everything from the clouds to the seasons, from the tides to the cycles of the moon. A heliocentric vision of our solar system underlies our general conceptions about the world; to delay its subsidiary power as an interpretative tool would be to deny a student his scientific heritage. Even if we were to withhold a discussion of the issue of Copernican vs. Ptolemaic theory while the student studied mathematics, the student would busily a-critically assimilate modern conceptions of the nature of mathematics in science as well as the subsidiary background necessary to employ any particular mathematical formula. No matter how we choose to educate future scientists, the inculcation of numerous presuppositions are essential. Introduction to science does not involve, for example, a single a-critical leap to the rationality of assessment, but rather, involves an initiation into a complex of basic assumptions about the referents, intentions and interpretative mechanisms of science itself.

Differing interpretations of these traditions therefore have significant educational bearing.

It is no accident that the controversy over evolution is centered in schools. The presuppositional implications of Darwinian evolution theory and Creationist theory are at issue over and above the specific evidence for or against each side; it is the presuppositional message that children assimilate in their studies at school. Although Darwin himself had a degree of belief in "supernatural beings" (e.g. ghosts), his theory of evolution seems to have been quite separate from such notions. Evolutionary theory is often thought to imply a Godless world. By implication, such a world seems to render itself without any intrinsic goal or purpose. Creationists see the world as having a primary intelligence which guides it toward a greater realization of some overriding purpose. Faced with similar information, Creationists see epiphany where Darwinians see random mutations; Darwinians see natural selection where Creationists see the hand of the divine.[13] The teacher of evolution assumes similar presuppositions in his account of evolution though no word of God or purposeless mechanism

---

[13]It does not matter if there is not a single word ontologically accurate in either Darwinian or Creationist theory; the contrast between the two clearly illustrates the significance of the presuppositional tenets of scientific inquiry. The presuppositions will not only shape the future direction of scientific research but also the relative value of the evidence eventually found.

be spoken. The presuppositional image of reality and science presented to the grade school student forms the idiom of his future conceptions. In presenting a metaphysical presupposition, whether it denies or affirms principles operating at higher levels than those directly observed in a given field, the teacher provides the student with a context for future inquiry.

Polanyi would not, I believe, present both sides of the evolution controversy so that students could decide for themselves. The interpretation of evidence is too sophisticated a process for novices just as we had seen in our discussion of the issues relating to the Copernican revolution. Rather, Polanyi would hope that the teacher exercise responsibile judgment for his position. The teacher should both affirm the presuppositional tenets of the scientific theory to which he subscribes as well as help students focus questions which may well lead to the dissolution of these presuppositions themselves. James Wagener explains that the authority of the teacher "is not the authority of conclusions but of the presuppositions of the intellectual heritage which she shares."[14] The student, having acquired the subsidiary use of the teacher's perspective may then reject it in the light of

---

[14] James W. Wagner, "The Policy of Michael Polanyi as a Source for Educational Theory," Ph.D. dissertation, The University of Texas at Austin, 1968, pp. 146, 147.

the conclusions he reaches upon further inquiry. The
freedom of the student to criticize and reject his
teacher's conclusions and commitments is essential to the
future viability of science. Science proceeds upon the
presuppositions its novices acquire through education;
it can, however, sufficate itself if teachers are not
seen as fallible in their judgments and presuppositions.
Critical reflection based upon given presuppositions
must counterbalance faith; scientific tradition is, in
essence, heuristic only; it is not prescriptive. Polanyi
(assuming ontological referents of science) states,

> Any tradition fostering the progress of
> thought must have this intention: to teach
> its current ideas as stages leading on to
> unknown truths which, when discovered, might
> dissent from the very teachings which en-
> gendered them.[15]

Yet other and perhaps more fundamental educational
implications arise from Polanyi's concept of tacit knowledge. We have previously seen that all assertions (scientific and otherwise) contain a-critical presuppositions.
These presuppositions are subsidiary rather than focal
as they function in the generation of rather than as the
content of assertions. Thus, to understand an assertion
requires one to acquire a subsidiary knowledge of its
presuppositional tenets. Such knowledge would enable

---

[15]Michael Polanyi, The Tacit Dimension (Garden City,
NY: Doubleday and Company, 1967), p. 82.

an individual to subsidiarily apply a given knowledge claim in new and yet unprecedented situations.

Let us recall our discussion of "latent learning." In Chapter III we discussed how rats familiar with a maze, when faced with new obstacles in their paths, unerringly devised new routes to their desired goals. The rats were able to apply their previous experiences in a subsidiary fashion to focus upon the new and varied problems they faced. Their subsidiary knowledge of the maze enabled them to focus on relations amongst varied channels that they had not heretofore had. The rats, as it were, subsidiarily anticipated changes in their routes.

It should not be assumed that the rats were performing what might be equivalent to deductive operations requiring no subsidiary knowledge. The rat's ability to walk down passages they had hitherto walked up, to make left turns where they had previously turned to the right, indicates that they were capable of comprehending the specifics of their previous experience and of anticipating unprecedented relations amongst the passages. The specific choices the rats made are less significant than the fact that such choices were contingent upon the rats' capacity to comprehend the very possibility of the varied use of specific passages.

The "latent" quality of learning whereby an individual is capable of anticipating diverse future experiences is a significant aspect of much human knowledge. To have such knowledge indicates that an individual has grasped an operational principle which transcends the particulars of previous experience. When we consider the use of symbols, as in human speech, the acquisition of the word refers to the assimilation of its subsidiary use beyond the limitations of past instances. The operational principle is predispositional; it is pre-linguistic; it underlies the generation of the word in the act of speaking. Proper use of a word implies that one has subsidiarily grasped its presuppositional foundations. More generally, Polanyi states,

> In learning to speak, every child accepts a culture constructed on the premises of the traditional interpretation of the universe rooted in the idiom of the group to which it was born, and every intellectual effort of the educated mind will be made within this frame of reference.[16]

More generally, the acquisition of such knowledge does not merely imply that when an individual is presented with a given stimulus, he can produce a _particular_ response; the acquisition of knowledge is not, as B.F. Skinner would argue, simply a change in behavior. The acquisition of knowledge implies that an individual can subsidiarily apply facts, concepts and general information in a variety of new and unanticipated contexts; it

---

[16]Polanyi, _Personal Knowledge_, p. 112.

implies that one has assumed a new intellectual dwelling place which predisposes one to respond to a new situation in a cohesive manner. Polanyi writes, "The interpretive framework of the educated mind is ever ready to meet somewhat novel experiences, and to deal with them in a somewhat novel manner."[17]

In essence, to acquire knowledge means that an individual has crossed a gap between the symbol and that which is symbolized, between the outer appearance of an object or word and its inner workings, between the previous instances of a term or concept and its heuristic functions.

The acquisition of knowledge implies a conceptual transformation; one's heuristic vision is altered to some degree and the interpretation of future events will differ accordingly. Relativity theory transforms one's concepts of space and time: physicists research problems unimaginable in a Newtonian universe and leave questions that might once have been essential. Freudian theory reshaped conceptions of human motivation and personality; Copernican theory displaced Man in the universe; a vast number of discoveries have shaped the questions asked and the problems pursued in medicine, astronomy, chemistry, psychology, etc. Each new discovery is based

---

[17]Polanyi, Personal Knowledge, p. 124.

upon the presuppositions provided by the discoveries preceding it, and provides, in its turn, the subsidiary context for new discoveries. The evolution of our intellectual heritage has not proceeded brick by brick like the construction of a building; it has emerged in constant transformation where each new knowledge claim transforms the knowledge which precedes it and shapes the course of future inquiry.

The point here is that, as each new addition to knowledge transforms our cultural presuppositional vision of reality, so each new lesson a student learns transforms his vision. In acquiring knowledge, one views the world differently, the relative significance of the problems one faces and the mode through which one addresses them. A predispositional shift occurs with each act of learning. The latent quality of the shift, the ability to reinterpret experiences and make new and varied situations comprehensible is present in the individual as a heuristic vision. It identifies new areas of inquiry and oversees future intellectual pursuits.

The educational implications of this single concept could reshape American educational goals and practices. It has been my experience that American education has been greatly influenced by the concept of knowledge implicit in behavioristic psychology. In the vast majority of teacher education courses and elementary school classrooms I have visited educators have increasingly emphasized the use of lesson plans utilizing specific behavioral objectives. These behavioral objectives indicate the precise behaviors students are expected to exhibit upon the completion of the lesson. The success of the lesson is said to be measured by the degree that the student acquires appropriate behavioral responses to specified stimuli. Learning is thus defined as a change in behavior, and it is thus assumed that a person having learned can exhibit specific behaviors.

This type of learning amounts to simply acquiring a given behavior when presented with a given stimulus or class of stimuli; it does not imply any latent relations with other acquired behaviors. Though stimuli may require a set of acquired behaviors, each of the behaviors is considered independent and as having no transformational effect on the others. One's knowledge is generally considered to be composed static behaviors which, though often in series, are isolated and distinct.

Polanyi, however, would argue that everything an individual learns transcends a particular behavior; it integrates dynamically with other concepts and predispositions to yield new perspective and insights. A young child having learned to count by two begins to group objects in pairs and to develop a notion of grouping, perhaps even symmetry. The third grader learning

grammar acquires a wider lesson that language has specific structures which can be identified and applied.
The sixth grader studying the ancient Roman respect for
law may well reassess his respect for the rules of his
schoolyard games. The eighth grader studying the cycles
of the seasons may generate new and varied inquiries
about the life cycles of birds or plants. The tenth
grader studying Darwin may reevaluate the significance
of his religious presuppositions. The senior studying
the works of Shakespeare may arrive at a variety of insights into his own aspirations and pains.

These examples by no means pertain only to what
educators often call the "affective domain." They are
the result of latent predispositional quality of knowledge.
Each fact learned raises new questions and directs us
to new conceptual structures. The coherences we perceive in the world around us assume new shape and dimension. Polanyi states, "Education is latent knowledge,
of which we are aware subsidiarily in our sense of
intellectual power based on this knowledge."[18]

The emphasis on direct behavioral changes provides
a clear and specific basis for measuring what behaviors
a student has acquired, but it distorts the measure
and understanding of his intellectual growth. The

---

[18]Polanyi, Personal Knowledge, p. 103.

intellectual impact of the acquisition of a given concept lies in its subsidiary functioning: its integration with and transformation of previous assumptions to generate new questions and insights. It functions like an operational principle at work in a given field of inquiry. Just as the operational principle transcends the particulars of a given set of empirical data and may be expected to reveal itself in new and unexpected ways in the future, so the subsidiary functioning of a concept transcends the particulars to which the concept once referred and can be expected to reveal itself in new and not necessarily logically deducible ways. The most significant indicator of one's knowledge is the general intellectual cohesiveness and scope of an individual's intellectual framework. The execution of particular behaviors has little value in this regard.

A student may well be able to exhibit an appropriate behavior when presented with a given response and yet be unable to apply this knowledge to varied stimuli. I recently heard a story of a young boy who after studying the Mississippi River in his class and performing quite well on his exam realized months later that the river he studied was the very same one that flowed a few miles from his house. He had acquired the desired behaviors but had been unable to use them to make the world around him comprehensible. On the other hand, it is quite possible for a child to forget the particulars of his lessons and yet retain the predispositions which they implied. Let us remember that the predispositions implicit in a given concept can, if rightly assimilated, acquire a subsidiary character which irreversibly changes the intellectual framework of the individual. Thus, a student may make great intellectual strides in that he may develop the ability to make that which is around him more rationally comprehensible while, at the same time, he may be unable to exhibit any specific and directly related behavioral changes. For instance, a student learning about the geography of a distant region might well be unable to identify the specific features of the area studied while he tacitly acquired the understanding that his own life is greatly affected by the nature of the terrain where he lives. Many adults, I am sure, have the same insight and have also forgotten the particular studies which provided them with such perspective.

It may well be that the American intellectual tradition is one which presumes a fragmented vision of knowledge and, corollarly, the world. The presuppositions which students have increasingly thrust at them direct students to see the world around them as fragmented and static. The subsidiary aspects of knowledge seem to play a decreasing role and the intuitive insights, the perspectives, the questions that might have been raised, go unnoticed. Perhaps the most devastating affect of the pervasive mechanistic concept of knowledge is that it well might cripple the development of subsidiary frameworks which provide the context for the particulars of one's intellect; we may be instructing students in the specifics of our own culture's knowledge while at the same time reducing the vision that could extend it. The lack of stimulation of the intuition and imagination to discern problems and advance inquiries (to seek new relations and possibilities) may well inhibit the creation of new and varied paradigms.

David Bohm, the eminent physicist, explains that,

> ...the idea of a separately and independently existent particle is...at best, an abstraction furnishing a valid approximation only in a certain limited domain. Ultimately, the entire universe...has to be understood as a single individual whole, in which analysis into separately and independently existing parts has no fundamental status.[19]

We can extend this notion to an image of human knowledge in which the notion of "separately and independently existent parts has no fundamental status." It is only in terms of an individual's subsidiary matrix that particular concepts or data receive their coherence. The notions that knowledge is merely a

_____

[19] David Bohm, Wholeness and the Implicate Order (London: Routledge and Regan Paul, 1980), p. 174.

change in behavior and can be objectively measured as such is misguided and intellectually inhibitive. The full measure of their implications for the future of science and for society may further emerge in the future.

The recognition of the subsidiary aspect of knowledge adds clarity to the development of curricula. When knowledge is viewed as mere behavior, the knowledge presented to the student can logically consist of any behavior that the student may reasonably be expected to acquire. Thus, a teacher could present a first grade class with a series of lessons on the planets expecting them to learn the features of the planets. However, once we recognize the subsidiary substructure of knowledge we may determine that such a subject is inappropriate for children with their limited presuppositional base. Although the students may be capable of naming the planets in the order of their distance from the sun and describe some of their features, the students probably have little presuppositional background to make the data they memorize capable of subsidiary use.

The information serves little, if any, function in expanding the intellectual framework of the student. The distances, temperatures and sizes have little subsidiary value to a child incapable of arriving at the question which these figures satisfy. The student does not know how to make the data integrate with his present conceptions of the world around him. The information is static and separate; it is unassimilable. The fact that Mercury is the closest planet to the sun makes little difference to a child viewing the night sky. To him, probably none of the stars seem close to the sun-- how could they be when it seems that the stars are out by night and the sun by day? The teacher may as well teach nonsense syllables for the words "closest," "furthest," "moons," "millions of miles" signify, in the end, absolutely nothing to the child.

Polanyi's concept of knowledge challenges the educator to devise curricula which extend from the child's heuristic vision of the world and bring it to bear upon questions and problems the teacher knows will help the student reveal new and wider insights of the world. Quasi-sophistication is unnecessary; the teacher ought only present the detail students need to satisfy and stimulate their desire for coherence given their pre-suppositional footing. As the teacher helps students focus in on new areas of concern and helps render them coherent to the degree that the student can subsidiarily apply them to other areas of interest, the student assimilates particular information and is apprenticed in the intellectual tradition of his society. Such an approach allows for the gradual emergence of the personally responsible knower within the confines of a given cultural heritage.

Yet, other implications of Polanyi's concept of personal knowledge can be found with respect to the role of the learner in the educational process. The context for the educational process, as we have already noted, consists in the learner's attempt to make increasingly wider and varied phenomena rationally comprehensible. It is a process in which the learner must be personally active in order to assimilate a concept and to understand the workings of its application. Thus, quite like John Dewey's pedadogical emphasis on "felt needs," Polanyi's model of personal knowledge requires that the learner arrive at problems which he believes will be a fruitful source of inquiry. In both cases, the student must personally identify a need and apply previous knowledge to its solution.

The major difference between the two perspectives lies in Dewey's emphasis on practical problems and scientific control and Polanyi's emphasis on the need for rational coherence and understanding. Dewey maintains that students must see practical purposes for their activities; these purposes often involve not only the individual but the community as a whole. The object of inquiry, therefore, is to "scientifically" arrive at a viable mode of action for solving common problems. Polanyi perceives the problems which spur inquiry as empty focuses which the individual attempts to clarify to his self-set, though culturally embedded, intellectual standards. The key here is not so much the practical mastery of a social problem but the personal intellectual mastery over a previously vague and partial aspect of reality. Dewey would be satisfied with a practical solution while Polanyi would maintain that the specifics of the given problem must be understood so that the principles at work there may be anticipated though not logically applicable to future areas of research.

In practical terms, the key words for Dewey would be "social relevance." Dewey might have a high school class set out to determine the social factors contributing to the use of drugs and the social consequences of drug use. Studies in a wide variety of disciplines from literature to science would focus upon finding practical solutions to the problems posed.

When we consider the same problem from Polanyi's perspective we will discover a large area of overlap with Dewey's concerns. Polanyi, in attempting to iden- tify the operational principles active in drug use, might study the same gamut of disciplines noted above. There is, in practice, a close relation between the search for practical control and rational comprehension. The practical difference lies in the fact that the latter, being a wider question, subsidiarily involves a larger scope of considerations. Polanyi's high school class could study the classics of Western Literature to reveal some of the factors which shape human motivation and purpose. Existentialist writers facing the question of the meaning or meaninglessness of existence could be used to widen the students' presuppositional foundations and challenge students to clarify their conceptions. A teacher could also use historical works exploring the struggles and goals of various figures in history as individuals exercising responsible judgment for their own existence within a cultural setting.

Polanyi states, "an admirer may be mistaken in choice of hero, but his relation to greatness is correct. We need reverence to perceive greatness even as we need a telescope to observe spiral nebulae."[20] The study of Charlemagne has little to do with the practical problems of our day which lead to drug use, but his conception and response to the troubles he perceived provide a context for the consideration of a single individual's responsible ordering of his own life. In any exemplar life the student chooses to study, we will see cultural, social, biological, and political factors feeding into the choices the individual makes with destiny. The student will see each person deciding for himself how

---

[20]Polanyi, Study of Man, p. 96.

much of the surrounding culture to accept as given and
making up his own mind what opportunities to take or
miss, what temptations to resist or succumb to. "Never
will the historian admit that such circumstances can
irresistibly determine a sane man's deliberate actions."[21]

Such studies yield perspective on one's own life;
they can form a subsidiary base for the integration of
the factors one addresses in one's own struggles. The
information acquired is not necessarily directly
applicable; it is not instrumental in the sense that one
could use the very same strategies and techniques as
(to continue our example) Charlemagne. The knowledge
gained functions subsidiarily to provide a wider, more
cohesive understanding of the subtle and complex
factors involved in the social, political, etc., stresses
that weigh upon one's own life. Such studies of literature and history pertain more to a personal quest to
make one's own life more rationally comprehensive than
an attempt to find a sheerly practical way to stem the
social problem of drug abuse.

We should not assume, however, that Polanyi would
want educators to limit the focus of the educational
process to practical problems. Just as Polanyi believes
. that scientific inquiry should proceed along those lines

---

[21]Polanyi, Study of Man, p. 89.

that scientists believe will reveal new rational coherences in nature, so it is reasonable to assume students should focus their efforts on finding new and varied patterns and relations in the subjects they study. No practical application of any particular knowledge discovered is necessary.

Perhaps the most important element of the problems considered by students in their studies is that they be thought-provoking; it is essential that they engender the student's personal activity to make sense of a given situation. In accordance with Polanyi's contention that coherences are found through an alternation of integration and analysis, an educator may stimulate a student's thinking by 1) providing clues to a focus which the student will have to integrate into a unity or 2) providing a general area of concern requiring students to identify and analyze specific components. It is assumed that, in either case, the task required is reasonably geared to the presuppositional foundations of the student.

It is important to remember that the processes of integration and analysis are dynamically related, neither method of provoking thought noted above requires only one of the two functions. Each method merely reflects one process or the other in a general fashion. Analysis and integration alternate in both cases.

Specific "thought provoking" problems of the first type might include having students seek number patterns in the multiplication tables, having them seek patterns in nature, e.g., spirals in pine cones, flowers, sea shells, treelimbs, etc., or having them copy a work of art placed upside-down in front of them. Each of these activities would provide the students with specifics which they have to integrate into a coherent unity.

Specific "thought provoking" problems of the second type might include having students describe an ocean to a person who had never seen one, having students map out the daily itinerary of an animal they have just studied, or having them analyze a current political struggle in terms of a character whom they had previously studied. All these activities would require the student to subsidiarily apply previous concepts to the analyses of familiar sets of data. These exercises may include but are not reducible to specific logical operations.

The essential element is, once again, that the student become personally active. In all cases the imagination is prodded to sketch new outlines and antici-pate the importance of given data, and the intuition is challenged to assess possibilities and outcomes according to presuppositionally grounded standards. The student thus learns to use these facilities, or at least to develop the conception that active thinking is essential to discovery. As a teacher I have often seen students who do not know how to extend lessons to new areas or root out the ideas underlying their reading selections. They rarely know how to arrive at questions; they memorize material without ever questioning its heuristic or presuppositional significance.

Further, we may also conclude the presuppositional nature of knowledge and the subsidiary aspect of knowledge acquisition imply that a liberal arts curriculum is of great value. Each subject provides a presuppositional concept of the principles operating in the environment, in society and the individual; each subject provides a subsidiary context that the student sees employed by those whom and under whom he studies. The liberal arts would provide the intellectual tenets necessary to make the wide and evasive questions of life and human purpose generally approachable. Thus, the liberal arts are what Harry Brody calls "value exemplars." He states, "In time, if schooling is successful, we dwell in these values; to use Polanyi's term, they are the probes by which we explore the domain of values."[22]

---

[22]Harry Brody, "Tacit Knowing as a Rational for Liberal Education," Teachers College Record, 80, 3 (February, 1979): 456.

Moreover, Polanyi's epistemology has implications for the student-teacher relationship. The teacher is the representative of tradition, of the primary assumptions, perspectives and values of the culture. He has a subsidiary grasp of such riduments that the student has yet to learn. Once again, the transmission is not accomplished through explicit instruction alone. For Polanyi, tradition is a presuppositional heritage that gives rise to, but exists beyond, explicit conceptual structures. To acquire explicit knowledge of a culture does not necessarily facilitate the use of the presuppositions that generate the specifics of the culture itself. Polanyi explains,

> This assimilation of great systems of articulate lore by novices of various grades is made possible only by a previous act of affiliation, by which the novice accepts apprenticeship to a community which cultivates this lore, appreciates its values and strives to act by its standards...the intellectual junior's craving to understand the doings and sayings of his intellectual superiors assumes that what they are doing and saying has a hidden meaning which, when discovered, will be found satisfying to some extent.[23]

Consequently, the teacher is like a master--because of his subsidiary facility with the tenets of culture and knowledge--while the student is like an apprentice-- because of his acquiring such a facility through continued

---

[23]Polanyi, Personal Knowledge, pp. 207-208.

association with the master. The apprentice learns to view the world, to make sense of all that surrounds him, through his observation of the master's subsidiary integrations. Thus, whatever the explicit structures may be, true education proceeds when the student learns to employ the heuristic framework undergirding his society and culture.

A further extension of this implication suggests that teaching requires a human teacher. The subsidiary use of knowledge is varied; it implies considerable emphasis on innovative but not capricious thinking. The personal judgment required in guiding a student to know the difference is perhaps more subtle and complex here, in the classroom, than it is in questions of scientific discovery and assessment.

More generally, we can conclude that Polanyi's model of knowledge necessitates that a learner be tacitly active in the educational process in order to make that which is studied a subsidiarily functioning item, an interpretative tool. The student is not a passive receptacle for information. He is an active individual who intervenes between the presentation of information and the responses he shapes. A major component of the educative process is the student's expansion of his own tacit facilities.

As we have seen, latent learning is a process by

which individuals make conceptual innovations; it is not a mere accumulation of specific behaviors. When one learns something, the knowledge gained becomes a new interpretative mechanism whereby antecedent knowledge may well be altered (or given new subsidiary powers) and previously unassimilable data may become rationally comprehensible.

The last educational implication of Polanyi's epistemology to be discussed extends beyond the parameters of education per se; it deals more generally with the whole question of human relations. It is clear that Polanyi rejects a behavioristic image of man. Just as Polanyi maintains that the proper referent of science transcends the particulars of observation, so a proper understanding of an individual transcends a mechanistic analysis of his behaviors.

We noted previously that the measure of one's intellectual development is not properly evaluated by the accumulation of separate and distinct facts and information. The latent quality of one's assimilation of a given subject, its heuristic function, resides in the individual more as a capacity yet to be tapped than a fulfilled potentiality. We may now add that understanding the mind of another (and specifically a student) requires that one not attend to specific behaviors of the individuals as things in themselves but as indicators of a deeper lawfulness or operational principle.

Such a concept is applicable to understanding the primary motives for learning. Polanyi maintains that the quest for rational coherence is primary to the inquisitiveness and enthusiasm for knowledge manifest by many intelligent organisms. Behavioristic conceptions of motivation which couple the acquisition of behaviors with particular reinforcers fail to recognize such an operational principle works within the individual to bring his environs under some sort of intellectual control. In order to understand what motivates a child one must read his behaviors as signs of attempts at establishing coherent conceptual structures.

The quest for coherence is not reducible to specific curriculum areas but encompasses the riddles of the child's own existence and that of the world around him. This does not lead the teacher to a psychologistic framework where given behaviors are said to be indicative of various affective mechanisms. The educator's goal would be to attempt to heed the principles at work within the child which lead him in a quest to make himself and the world comprehensible.

In the final analysis, Polanyi's theory of knowledge implies that education is a process whereby members of society are initiated into its presuppositional perspectives. Through education, one acquires more than a mass of facts and skills; one acquires a vision of humanity and nature according to which he conceives of himself and his goals.

It may be concluded that education provides a heuristic basis for the development of an individual's morals and ethics. Unlike Dewey, Polanyi asserts that societal presuppositions yield the primary context for attending to humanity and nature. This context, rather than the specific findings of science, provide individuals with a fiduciary rather than wholly critical basis for value and action. Thus, education in its widest sense is a process whose goal is the development of moral character.

BIBLIOGRAPHY

Agassi, Joseph. "Genius in Science." Philosophy of the Social Sciences 5 (June 1975):49-60/

_____. "Sociologism in Philosophy of Science." Metaphilosophy 3 (April 1972):103-122.

_____. "Rationality and the 'Tu Quoque' Argument." Inquiry 16 (Winter 1973:395-406.

_____. "When Should We Ignore Evidence in Favor of a Hypothesis?" Ratio 15 (December 1973):183-205.

Brennan, John. "Polanyi's Transcendence of the Distinction Between Objectivity and Subjectivity as Applied to the Philosophy of Science." Journal of the British Society for Phenomenology 8, 3 (October 1977):141-152.

Brody, Harry S. "On 'Knowing With.'" Proceedings of the Education Society 26 (1970):89-103.

_____. "Tacit Knowing as a Rationale for Liberal Education." Teachers College Record 80, 3 (February 1979):50-66.

Bohm, David. "On Insights and Its Significance for Science, Education and Values." Teachers College Record 80, 3 (1979):403-418.

_____. Wholeness and the Implicable Order. London: Routledge & Kegan Paul, 1980.

Brodbeck, Mary. "A Review of Personal Knowledge." American Sociological Review XXV (August 1960):582-583.

Brownhills, R.J. "Political Education in Michael Polanyi's Theory of Education." Education Theory 23 (Fall 1973):303-309.

Buber, Martin. Between Man and Man. Trans. Ronald Gregor Smith. New York: The Macmillan Company, 1968.

Burtt, Edwin A. The Metaphysical Foundations of Modern Science. Garden City, NY: Doubleday & Company, 1932.

Clark, Ronald W. Einstein: The Life and Times. New York: New World Publishing Co., 1971.

Cohen, Robert S., and Joseph Agassi. "Dinosaurs and Horses, or: Ways with Nature." Synthese 32 (November-December 1975):233-247.

Dewey, John. Democracy and Education. New York: The Macmillan Company, 1916.

_____. Experience and Nature. Open Court: La Salle, 1971.

_____. The Child and the Curriculum and The School and Society. Chicago: The University of Chicago Press, 1968.

Diller, Ann. "On Tacit Knowing and Apprenticeship." Educational Philosophy and Theory 7 (March 1975): 55-63.

Gill, Jerry H. "The Case for Tacit Knowledge." Southern Journal of Philosophy 9 (Spring 1971).

Grene, Marjorie, ed. The Anatomy of Knowledge. Amherst: The University of Massachusetts Press, 1969.

_____, ed. Intellect and Hope. Durham, NC: Duke University Press, 1968.

_____, ed. Interpretations of Life and Mind. New York: Humanities Press, 1971.

_____. . The Knower and the Known. Berkeley: University of California Press, 1974.

_____, ed. The Logic of Personal Knowledge: Essays Presented to Michael Polanyi on His Seventieth Birthday. Glencoe: The Free Press, 1961.

_____. "Tacit Knowing: Grounds for a Revolution in Philosophy." Journal of the British Society for Phenomenology 8, 3 (October 1977):164-171.

Hall, Ronald Lavon. "The Structure of Inquiry: A Study in the Thought of Michael Polanyi." Unpublished Ph.D. dissertation. Chapel Hill: University of North Carolina, 1973.

261

Harre, Rom. "The Structure of Tacit Knowledge."
     Journal of the British Society of Phenomenology 8,
     3 (October 1977):172-177.

Hume, David. A Treatise of Human Nature. New York:
     P.L. Dutton and Co., 1911.

Kant, Imanuel. Critique of Pure Reason. Trans. Norman
     kemp-Smith. New York: St. Martin's Press, 1965.

Koestler, Arthur. The Sleepwalkers. New York:
     Grosset and Dunlap, 1959.

Kroger, Joseph. Polanyi and Lonergan on Scientific
     Method." Philosophy Today XXL, 144 (Spring 1977):
     2-20.

Kuhn, Thomas S. The Structure of Scientific Revolutions.
     Chicago: The University of Chicago Press, 1970.

Maslow, Abraham H. The Psychology of Science. New
     York: Harper and Row, 1966.

Piaget, Jean. Psychology of Intelligence. Trans.
     Malcolm Piercy and D. Ederlyne. London: Routledge
     & Kegan Paul, 1950.

Poincare, Henri. Scientific Method. Trans. Francis
     Martland. New York: Dover Publications, 1952.

Polanyi, Michael. "This Age of Discovery." The
     Twentieth Century CLIX, 949 (March 1956):227-234.

_____. "The Creative Imagination." Tri-Quarterly
     8 (Winter 1967):111-123.

_____. "Faith and Reason." The Journal of Religion
     LXI, 4 (October 1961):237-247.

_____. "From Copernicus to Einstein."
     Encounter V, 3(September 1955):54-63.

_____. "Logic and Psychology." American
     Psychologist 23, 1 (January 1968):27-43.

_____. The Logic of Liberty: Reflections and
     Rejoinders. Chicago: The University of Chicago
     Press, 1951.

_____. "On Body and Mind." The New Scholastic
     XLIII, 2 (Spring 1969):195-204.

Polanyi, Michael. _Personal Knowledge:_ _Towards a Post-_
      Critical Philosophy. New York: Harper and Row,
      1964.

_____. "Problem Solving." _The British Journal_
      for the Philosophy of Science VIII (30 August
      1957):89-103.

_____. "Science and Conscience." _Religion in Life_
      XXIII, 1 (Winter 1953-54):47-58.

_____. "Science and Reality." _The British Journal_
      for the Philosophy of Science 18 (1967):177-196.

_____. _Science, Faith and Society_. Chicago:
      The Chicago University Press, 1964.

_____. "Science Outlook: Its Sickness and Cure."
      Science 125 (15 March 1957):480-504.

_____. "Scientific Beliefs." _Ethics_ LXI, 7
      (October 1950).

_____. _The Study of Man_. Chicago: The University
      of Chicago Press, 1958.

_____. _The Tacit Dimension_. Garden City: Doubleday
      & Company, 1967.

_____. "Words, Conceptions and Science." _The_
      Twentieth Century CLV, III (Spring 1955): 256-267.

Polanyi, Michael, and Harry Prosch. _Meaning_. Chicago:
      The University of Chicago Press, 1975.

Popper, Karl. _Conjectures and Refutations_. New York:
      Basic Books, 1962.

_____. _The Logic of Scientific Discovery_. London:
      Hutchinson, 1959.

_____. _The Open Society and Its Enemies_. New
      Jersey: Princeton University Press, 1966. Vol. II.

_____. _Objective Knowledge_. Oxford: The Claren-
      don Press, 1972.

Prosch, Harry. "Biology and Behaviorism in Polanyi."
      Journal of the British Society for Phenomenology
      8, 3 (October 1977):178-191.

Prosch, Harry. "Polanyi's Tacit Knowing in the 'Classic' Philosophers." Journal of the British Society for Phenomenology 4, 3 (October 1973):201-216.

Rogers, Carl R. "On Our Science of Man." Man and the Science of Man. Columbus, OH: Charles E. Merrill, 1968, pp. 52-72.

Ryle, Gilbert. The Concept of Mind. New York: Barnes and Noble, 1967.

Scott, William T. "Commitment: A Polanyi View." Journal of the British Society for Phenomenology 8, 3 (October 1977):192-206.

Scott, William T. "Polanyi's Theory of Personal Knowledge: A Gestalt Approach." The Massachusetts Review (Winter 1962):349-368.

_____. "Tacit Knowledge and the Concept of Mind." Philosophical Quarterly 21 (January 1971):22-35.

Tiedeman, Kent H. "Personal Knowledge: The Epistemology of Michael Polanyi." Unpublished Ph.D. dissertation. Cincinnati: University of Cincinnati, 1971.

Wagener, James W. The Philosophy of Michael Polanyi as a Source for Educational Theory." Unpublished Ph.D. dissertation. Austin: University of Texas, 1968.

_____. "Toward a Heuristic Theory of Instruction: Notes on the Thought of Michael Polanyi." Educational Theory 20 (Winter 1970):46-53.

B 945 .P584 K36 1984

Kane, Jeffrey,

Beyond empiricism

A113 036168b 3